Polymer-Silica Based Composites in Sustainable Construction

Polymer-Silica Based Composites in Sustainable Construction

Theory, Preparation and Characterizations

Harrison Shagwira
Fredrick Madaraka Mwema
Thomas Ochuku Mbuya

CRC Press is an imprint of the
Taylor & Francis Group, an **informa** business

First edition published 2022
by CRC Press
6000 Broken Sound Parkway NW, Suite 300, Boca Raton, FL 33487-2742

and by CRC Press
2 Park Square, Milton Park, Abingdon, Oxon, OX14 4RN

CRC Press is an imprint of Taylor & Francis Group, LLC

Library of Congress Cataloging-in-Publication Data
A catalog record has been requested for this book

ISBN: 978-1-032-14011-7 (hbk)
ISBN: 978-1-032-14012-4 (pbk)
ISBN: 978-1-003-23193-6 (ebk)

DOI: 10.1201/9781003231936

Typeset in Times
by Deanta Global Publishing Services, Chennai, India

To my parents, John and Martha Shagwira; you have been such an inspiration in my life. Thanks for your love and support.

Harrison Shagwira

I dedicate this book to the loving memory of my dad and elder sister, sleep on well. To my wife, Mwihaki, and twin sons, Mwema and Nyika, thanks for love and support.

Fredrick M Mwema

Contents

Foreword

I am delighted to write this foreword, not only because the authors and I have been friends and colleagues for more than four years, but also because I believe in the research value of their work in materials and especially on the new trends of materials and polymers. I also believe that the research at every level and stage should intend and purpose to improve and make the world a better place. This can be done by the reduction of emissions and the invention of new methods of doing things.

In Chapter 1 of this book, the authors have shown how green composite materials can be made using agricultural wastes like rice husks, banana peels, and animal caucuses. This shows how much the research in the materials especially polymers has established its roots in the industry. This book provides high professional insights on polymer materials in the construction industry. Construction industries should adapt to this knowledge and research in this book to improve the industry and aesthetics of buildings and other constructions throughout the world. The authors produce powerful insights on the development of materials towards Industry 4.0. The materials and systems described in this book also ensure the reduction of emissions and increase the recyclability of plastics and other debris in quarries.

Reading this book, you will find how recycling and the use of 3D printing elements will improve living standards. It will give you insights into the efficiencies of polymers and 3D printers. The authors provide compelling evidence on polymers and their use in different aspects of engineering.

I hope that this book will become a primer for researchers and engineers on their material advancements, knowledge, and also in the implementation of Industry 4.0 concept.

Patrick W. Kariuki
Dedan Kimathi University of Technology, Nyeri, Kenya

Preface

Polymer-based composites have been widely adopted by various industries due to the attractive attributes they offer. In construction industry, these composites have been cited to have the potential to enhance sustainability and advancement. There are concerted efforts to utilize waste materials in the construction sector to enhance recycling and reduce environmental pollution. As it is known, polymers, especially thermoplastics, are the major pollutants of our environment. The construction industry, today, is adopting the use of waste polymer-based composites for its structural and aesthetic applications. As such, there is a lot of research and innovation from the scientific and engineering communities to develop a wide range of composites from waste plastic and reinforcements for construction applications.

This book contributes to this subject by providing insights into the progress in composite materials, with a focus on plastic-silica composites, for green construction industry. The book is divided into six chapters. Chapter 1 introduces composite materials for the eco-friendly era and classification and application of polymer composites. Chapter 2 presents processing, testing, and failure modes of composite materials. In Chapters 3 and 4, the life cycle analysis of plastics, pollution, and utilization of plastic and silica waste are discussed. In Chapter 5, a typical polymer-silica composite from high density polyethylene/polypropylene, herein abbreviated as HDPE/PP-quarry dust is described with the aim of illustrating the processing of these composites for the construction industry. Finally, in Chapter 6, an outlook of the future of composite materials for construction sector is discussed with an emphasis on the role of Industry 4.0 in the sector. The book will be of help to researchers, students, and engineers in polymer-based composite field and those interested in developing materials for sustainable construction industry.

Harrison Shagwira ***Dedan Kimathi University of Technology, Nyeri, Kenya***

Fredrick M Mwema ***Dedan Kimathi University of Technology, Nyeri, Kenya & University of Johannesburg, Auckland Park, South Africa***

Thomas O Mbuya ***University of Nairobi, Nairobi, Kenya***

June 2021

Acknowledgements

The authors of the book acknowledge Dedan Kimathi University of Technology and the University of Nairobi for providing research platforms to them. The authors also acknowledge TH Wildau staff under Professor Michael Herzog for allowing access to the polymer science laboratory. Deutscher Akademischer Austauschdienst (DAAD) is also acknowledged for the sponsorship of the first author (Harrison Shagwira) to travel and stay in Germany during the research. National Research Fund (NRF-Kenya) is also acknowledged for sponsoring the project on green building materials for the construction sector, out of which this work was initiated.

The authors also acknowledge their colleagues and families for their support in the development and writing of this monograph. In a special way, Mr. Patrick Wanjiru Kariuki is acknowledged for proofreading the manuscript of the book.

Author Biographies

Harrison Shagwira

Harrison is a design engineer at Dedan Kimathi University of Technology (DeKUT) in the Department of Mechanical Engineering. He holds master's and bachelor's degrees, 2021 and 2017, respectively, in mechanical engineering from Dedan Kimathi University of Technology, Nyeri, Kenya. He has research interest in materials and engineering, specifically in polymer-composite processing, severe plastic deformation, and conventional material forming and processing. He has published seven journal articles, four conference articles, and two book chapters. He is a registered graduate engineer with Engineers Board of Kenya (EBK).

Fredrick Madaraka Mwema

Dr. Mwema is a lecturer at Dedan Kimathi University of Technology. He is also a postdoctoral researcher at the University of Johannesburg, South Africa. Currently, he is Chair of the Department of Mechanical Engineering at DeKUT. He obtained BSc and MSc degrees in mechanical engineering from Jomo Kenyatta University of Agriculture and Technology (JKUAT), Kenya, in 2011 and 2015, respectively. He has a PhD in mechanical engineering from the University of Johannesburg, which he obtained in 2019. His PhD research work involved thin film coatings for surface protection and functional components.

He has interests in advanced manufacturing, severe plastic deformation processes, additive manufacturing, thin film depositions, surface engineering, and materials characterizations. In thin films, Dr. Mwema has interest in fractal theory of coatings for enhanced depositions and behaviour in advanced applications. He has published more than 70 articles in peer-reviewed journals, conferences, and book chapters. He has written two book monographs, published in 2020 and 2021. Dr. Mwema has contributed extensively to research on thin films and manufacturing with a Scopus H-Index of 9 and has over 200 citations. He supervises and mentors several students, currently with over four masters and PhD students. He has over six years of experience in teaching and training undergraduate students in mechanical engineering. Dr. Mwema is very passionate about local manufacturing in developing world, with focus in Kenya. He is currently championing for manufacturing and consumption of local products including construction materials through recycled materials and locally fabricated machines.

Thomas Ochuku Mbuya

Dr. Mbuya is a senior lecturer at the Department of Mechanical and Manufacturing Engineering at the University of Nairobi. In 2012, he obtained his PhD in engineering materials from the University of Southampton in the United Kingdom, and later in 2018 spent one year at the same university as a Commonwealth Rutherford Fellow. He has published extensively on microstructural characterization of materials and the micromechanisms of failure. His current research interest is in the development of sustainable models for materials resource utilization including recycling of wastes to new material products and generation of clean energy from agricultural waste.

1 Introduction to Composite Materials for Green Construction Industry

1.1 BACKGROUND

In today's world, manufacturers have started to focus on producing eco-friendly composite materials. New materials are being developed often to meet the demand for materials that possess excellent properties. However, as much as researchers are coming up with new materials, the big question is: How sustainable are these new materials? It is therefore food for thought for researchers to develop green materials, among them green composite materials. The term "green composite" refers to materials that are produced by combining other different components to enhance environmental and economic sustainability. The green composite materials tend to highly benefit the environment and they have found their applications in the construction industry [1].

For a better tomorrow, it is the responsibility of all sectors of the economy o commit to providing green materials. It is through such commitments that he manufacturing practices in today's world are slowly changing, hence we an see products made from materials such as: carbon fibre, Kevlar, and fibrelass [2]. Such positive practices not only protect the environment but also ncrease the sales of products in any particular business set up. The increase n the sales of products is attributed to the urge of the customers to assist in

DOI: 10.1201/9781003231936-1

environmental protection through the understanding of the risks posed to the environment through non-green manufacturing. Therefore, the manufacture of the products using green composite materials is a win-win for both the environment and the companies.

In addition, green composites can be produced from agricultural waste products [3]. Such agricultural wastes include: animal carcasses, crop residues, banana peels, husks, bones, hoofs, feathers, and bedding/litter [4]. Figure 1.1 shows agricultural wastes that can be used in the production of green composites. For these agricultural wastes to be utilized in the production of green composites for construction applications, the removal of impurities and undesired particles from agricultural wastes is necessary during the processing. During the production of green composites, these agricultural wastes can be used as the matrix or the reinforcement material based on the type of

FIGURE 1.1 Agricultural wastes that can be used in the production of green composites with (a) rice husks, (b) banana peels, (c) vegetable wastes, and (d) fruit wastes [5] (obtained for free from www.pinterest.com).

agricultural waste. Additionally, naturally occurring inorganic materials can also be used in the production of green composites. Such materials include and are not limited to quarry dust, soil, flint, glass, obsidian, granite, sand, gems, clay, sandstone, and plasticine.

Green composites (from natural inorganic materials and agricultural wastes) can in different ways be used in civil engineering for structural and non-structural applications, for example, in the construction of roofs, walls, pavements, partition walls, tiles, and embankments. The organic nature of agricultural waste-based fibres necessitates the need for developing new techniques that may delay or prevent the decay of the wastes in the green composites. Consequently, the connection between the structural performance of the green composites and the corresponding material characteristics is an aspect that needs to be investigated in detail.

1.2 COMPOSITES: MATERIALS FOR ECO-FRIENDLY CONSTRUCTION

1.2.1 Introduction

The construction industry is recognized to be one of the highest consumers of energy and materials while at the same time a major source of pollution [6]. The whole life cycle of construction has a crucial environmental impact, ranging from raw materials extraction to recycling and disposal. Material production/processing is a key factor in the life cycle of construction that has a negative environmental effect. The construction industry has therefore progressed tremendously towards meeting some of the key objectives of the sustainable development concept. Civil and structural engineers play a significant role in the reduction of the impact of new or existing construction methods on the environment. Some sustainable development objectives can be met through the use of specific technologies or other civil engineering systems.

1.2.2 How Construction Polymer Composites Differ from the Traditional Construction Materials

More than 30% of all produced polymers are used in the construction industry each year [7]. They provide multiple benefits over conventional materials, such as ease of processing, corrosion resilience, and lightness. Polymers can be mixed with additives/reinforcements to produce composites that have better properties so that they can be used as structural components. Polymer composites find their applications in various forms from the construction industry to space satellite and aerospace industries with high-tech applications. In recent years, polymer composites have had considerable advancements in their material properties to suit the construction sector.

The design procedures for polymer composite materials used in construction applications necessitate a broader understanding of the material and a greater research and development effort as compared to traditional construction materials. The properties associated with the composite produced are due to the process of design that takes into account several factors characterized by the material's anisotropic behaviour, covering stability properties, elasticity, micro-mechanical, and strength. These material properties are affected by environmental exposure, manufacturing techniques and loading history. Therefore, the design of composites involves designing the individual materials incorporated in the polymer matrix, designing the final composite produced, and the ability to understand the production process to be used.

In the recent past, construction practices such as bridge design, bridge repair, structural strengthening, and so forth have been using polymer composites. The application of polymer composites to construction activities is because of their considerably superior properties as compared to earlier composites, making them be referred to as advanced composites. Advanced polymer composites contain high-volume fractions of fibres with high stiffness and strength. Some of the additives to these polymer composites enhance properties like chemicals, fire and heat resistance, and properties which may not be present in conventional materials such as wood. Additionally, automated techniques have enabled the production volumes to increase hence lowering the costs of both resin and fibre components, making polymer composite compete with conventional building materials. This has caused a shift to the use of polymer composite in construction applications. Figure 1.2 shows a comparison between a wall constructed using polymer composite and a wall constructed using traditional materials. Clearly, walls constructed from polymer composites can be constructed faster and are more aesthetically appealing.

1.2.3 Repair and Rehabilitation of Polymer Composites Structures

Polymer composite materials are used in the rehabilitation of structures or used in the construction of new structures, and they enhance environmental

FIGURE 1.2 Showing a comparison between wall construction using (a) polymer composite, which is aesthetically appealing with ease of construction, and (b) traditional materials, which is poorly finished and takes a lot of time to construct (obtained for free from www.pinterest.com) [5].

sustainability by using the minimum amount of material resources and increasing the life span of structures. There is a worldwide increase in the number of infrastructures that require rehabilitation, upgrading, reinforcement, and/or repair. Figure 1.3 shows a damaged bridge that has been rehabilitated using fibre-reinforced polymer (FRP) composite. The conventional rehabilitation methods are affected by the emergence of new materials and techniques that provide professionals with convenient and affordable practices. Due to the advantages of polymer composites over steel and other traditional materials, new advances achieved with polymer composites have enabled engineers to overcome the challenges related to conventional rehabilitation techniques.

Research and development of new materials and techniques have been on the rise and most materials have been adopted in the construction industry's rehabilitation system. However, a number of construction stakeholders are reluctant to fully adopt these materials. The major reasons for this reluctance are: lack of design and detailing guidelines, code of practice, standards, and the absence of a definite structural performance of the composites during loading. The cheaply acquired steel materials may make the cost of construction polymer composites appear elevated. However, in the long run, the ease of storage, handling, installation, transportation, and the life cycle cost of polymer composites make their overall cost be less than that of steels.

The rehabilitation process is carried out through visual inspection, distress mapping, non-destructive tests, and so forth, to identify the reason why

FIGURE 1.3 Use of FRP composite in the rehabilitation of a bridge deck [8] (reused from Karbhari (2006)) (with permission from Elsevier Ltd).

damage has occurred on the structure. During rehabilitation, the process of strengthening can be either passive or active strengthening. In the active strengthening system, there is the introduction of external forces that counteract the internal forces that tend to damage a structure or a member. An example of active strengthening systems is using prestressed polymer composites laminates to bond structural members. The passive strengthening system is strengthening with no addition of external forces to the structure or member [9]. Examples of passive strengthening include wrapping polymer composites on damaged structures, bonding steel plates, and concrete jacketing.

1.2.4 Potential Application of Composites Repair and Rehabilitation Systems

Even at early stages after construction, structures made of concrete show signs of distress and deterioration due to factors such as physical and chemical changes, structural failure, and wear and tear. Consequently, it is inevitable to have a concrete repair and rehabilitation industry of equal importance as the construction industry. Composite materials, particularly the FRPs are considered as materials that can be used to strengthen concrete structure due to their high strength-to-weight ratio while making minimal additional weight to structures. FRPs are also used in strengthening masonry structures through an integrated system that utilizes epoxy resins and fibres such as glass, carbon, and aramid. This technique is ideal because it incorporates the advantages of utilizing advanced lightweight and non-corrosive FRP composites. Additionally, laminates become more efficient in making structures stronger by using the tensile capacity of prestressed carbon fibre-reinforced polymer (CFRP).

1.2.5 Codes and Standards in Pertaining Construction Materials

The term "standards" in the construction industry generally means published documents used in the definition of common procedures, methods, and specifications applied in construction. The importance of setting up standard guidelines is that the reliability and consistency of a certain material, product, service, etc. is guaranteed. Additionally, specifications generally refer to a series of material standards used, manufacturing quality, type of tests to be performed, etc. There exist several organizations that set standards for construction activities, some include [10]:

(i) The British standards: They are contained in the publication of standards made by the British Standards Institution which also is mandated to make and provide the public with tools for self-assessment, a variety of books, workshops, and conferences. Publications by the British Standard Institution are technical practices that may be used as guidelines for the product manufacture, evaluation of a process, or provision of a service.

(ii) The International Organization for Standardization (ISO): This is a non-governmental organization accountable for producing documents that contain guidance, requirements, characteristics, or specifications to be used in various fields such as the construction industry and ensure that there is consistency in processes, products, services, and materials.

(iii) The National House Building Council (NHBC): This organization is based in the UK and is responsible for providing warranties and insurance for homes that have been newly built. NHBC standards primarily focus on the houses' designs and their construction, preventing the construction of substandard houses.

(iv) The International Ethics Standards Coalition (IES): This is a group that makes publications on the ethics codes for stakeholders in infrastructure, construction, property, land, and similar professions.

(v) The International Property Measurement Standards Coalition (IPMSC): This is a coalition of nonprofitable organizations and professionals working together to develop additional international standards that can be used in property measurement. Such properties are like buildings specifically for global corporations that focus on cross-border property investment.

(vi) Other standards include and not limited to:
- International Construction Measurement Standards (ICMS).
- The Common Minimum Standards (CMS).
- Minimum space standards.
- Building control performance standards.
- Professional standards for architects.
- Lift Standards: EN 81-20 and EN 81-50.
- International Property Measurement Standards.
- Technical housing standards.
- Limiting fabric standards.
- Housing standards review.
- Common minimum standards.
- Scottish building standards.
- Draft housing standards.
- Minimum Energy Efficiency Standard (MEES).

1.3 CLASSIFICATION OF POLYMERS COMPOSITE MATERIALS USED IN CONSTRUCTION

1.3.1 Fibre-Reinforced Polymer Composites

Since the 1960s, there have been tremendous efforts towards the adoption of fibre-reinforced polymer (FRP) composites to the construction industry [11]. FRP composites have found applications in construction and are used as pressure pipes, tank liners, infill panels, load-bearing panels, roofs, and so forth. FRP composites are not only used as parts of a structure but can form the whole structure. For instance, FPRs are used in the construction of rail, road, and footbridges. The appealing aesthetic value of FRPs makes them also be used in fittings, cladding, and linings.

Major reasons why polymer composites are used in construction are:

- Time-saving, especially for projects which are under tight schedules due to their low weight, easing the construction process.

- Radio transparency.
- Durable and can withstand harsh conditions.
- Allows repair of structures to be carried out while still on site.
- Strengthening of structures can be carried out on site.
- Its anisotropic nature allows properties to be customized depending on the direction where high performance is required.
- Aesthetically appealing – texture, colour, or shape can be customized.
- Resistance to fire.
- Their low maintenance makes them suitable to be used in difficult to maintain conditions.

Some of the FRPs include:

(i) Carbon fibre-reinforced polymers

Nowadays, it is the desire of customers to know the manufacturing process carbon footprint and products' carbon footprint before they make any purchase. For instance, the automotive industry has had a massive development and utilization of structural carbon fibres [12]. This resulted in the emergence of greener cars in the market which have a higher market as compared to cars that use hydrocarbon fuels due to the use of carbon fibres and electricity to mention but a few. Green cars are usually lighter and use smaller engines that are more efficient and are associated with fewer emissions. The same approach is adopted by the construction industries to make buildings and structures green. Green buildings and structures will tend to use lighter materials, be easy to construct and decorate, and at the same time conserve the environment. Main applications of CFRP in construction industries include strengthening of steel bridges and structural members, construction of earthquake-resistant buildings and structures, rehabilitation of reinforced concrete and masonry columns, and control of corrosion in construction.

(ii) Glass fibre-reinforced polymers (GFRPs)

These are a type of polymer composites that are produced by incorporating glass fibre into a polymer matrix. These composites are mainly produced by moulding and setting. GFRPs are well known for their decorative applications due to their great light transmittance capabilities. Some other appealing properties include corrosion resistance, nonmagnetic, lightweight, high strength, fatigue resistance, and moisture resistance. The general applications of GFRP in the construction industry are: interior decoration and construction of handrail systems tanks, platforms, ladders, pipe, and pump support.

(iii) Wood fibre-reinforced polymers

In these polymer composites, wood fibres are added to a polymer matrix to enhance its properties. They are usually characterized by strength-to-weight ratio, high damping property, durability, flexural strength and stiffness, and resistance to impact, wear, and corrosion. The applications of these composites in construction industries include: construction of floors, ceilings, walls, pavements, and partition walls.

1.3.2 Particle-Reinforced Polymer Composites

These are a type of polymer composites that use particles as reinforcing additives in the polymer matrix. Unlike the anisotropic behaviour of fibre-reinforced composites, the material properties of particle-reinforced composites are affected by the amount, shape, and type and distribution of the reinforcing particles. They are usually applied where there is a high need for wear resistance, e.g. on the floors. Some of the particles used in producing these composites include: metal powder, granular granite, quarry dust [13], marble powder, basalt powder, concrete, silicon carbide, and silica sand. The application of particle-reinforced polymers in construction include: production of roofing tiles, walls, floor tiles, pavements, roads, and bridge surfaces.

1.4 APPLICATION OF POLYMER COMPOSITES IN CONSTRUCTION

1.4.1 Roofing Tiles

Polymer composite can be used in the production of roofing tiles. Roofing tiles are exposed to harsh conditions such as exposure to ultraviolet radiations, temperature variations, wet and humid conditions, and loading. Such conditions may cause weakening and breakage of the roofing tiles. For instance, temperature variations may cause thermal stresses and fatigue to the roofing tiles. Therefore, roofing tile materials should be heat resistant. Other desirable properties roofing tile should possess are corrosion resistance, high strength, durability, and should resist colour change. Polymer composites such as plastic-quarry dust composites possess such properties. Figure 1.4 shows a typical application of polymer composites in roof construction.

FIGURE 1.4 Durables, lightweight, and eco-friendly polymer composite roof tile with aesthetically pleasing slate appearance (obtained for free from www.pinterest.com) [5].

1.4.2 Surface Applications

Surface applications include cladding, construction of floors, roads, and pavements. These surfaces are exposed to chemicals, heat, heavy loads, scratching, and so forth. Therefore, the applications require materials that are heat resistant, wear-resistant, hard, corrosion-resistant, high strength, and relatively light to ease the transportation and construction process. Polymer composites such as concrete-polymer composites can be well suited for these applications. A typical application of polymer composite in the construction of floors and pavements is shown in Figure 1.5.

1.4.3 Construction of Structures and Members

Polymer composites can be used in the construction of members of a structure, or the whole structure. Engineers are focusing on using light materials ith high strength. This is meant to reduce unnecessary weight (loading) to e structure members. Some of the structures that utilize polymer compos-ites include: walls of buildings, full-bridge structures, bridge decks, bridge

FIGURE 1.5 Showing different applications of polymer composite in construction with (a) as a floor tile and (b) in pavements (obtained for free from www.pinterest.com) [5].

FIGURE 1.6 Bridge made up of GFRP composite under construction in Friedberg, Germany. Image courtesy of Fiberline, Composites A/S [14] (reused with permission from Elsevier Ltd).

enclosures, masts and towers, water control structures, modular structures, domes structures, and architectural mouldings. Figure 1.6 shows a bridge structure that is made up of polymer composite.

Other applications of polymer composites in construction industries include: interior decoration; construction of ceilings, pipes, tanks, access cover and sanitary ware; strengthening of existing structures; and seismic retrofitting

1.5 SUMMARY

In this chapter, an introduction to green polymer composites has been made with the focus being on the construction industry. The general types of polymer composites have been outlined. Distinctions between construction polymer composites and traditional construction materials have been discussed. These polymer composites have been shown to find a wider application in the construction industry ranging from construction of structures, cladding, rehabilitation, and repair of structures to production of roofing tiles, floor tiles and pavements, etc. Clearly, the future of construction lies in advanced green polymer composites.

REFERENCES

1. K. Singha, P. Pandit, S. Maity, A. Ray, and V. Kumar, "Advanced applications of green materials in construction applications," in *Applications of Advanced Green Materials*, S. Ahmed, Ed. Sawston, UK: Woodhead Publishing, 2021, pp. 223–238.
2. H. Fouad, A.-H. I. Mourad, B. A. ALshammari, M. K. Hassan, M. Y. Abdallah, and M. Hashem, "Fracture toughness, vibration modal analysis and viscoelastic behavior of Kevlar, glass, and carbon fiber/epoxy composites for dental-post applications," *Journal of the Mechanical Behavior of Biomedical Materials*, vol. 101, p. 103456, 2020, doi: 10.1016/j.jmbbm.2019.103456.
3. N. P. Vignesh, K. Mahendran, and N. Arunachelam, "Effects of industrial and agricultural wastes on mud blocks using geopolymer," *Advances in Civil Engineering*, vol. 2020, p. 1054176, 2020, doi: 10.1155/2020/1054176.
4. H. Shagwira and F. M. Mwema, "Advances in animal/plant–plastic composites: preparation, characterization and applications," in *Plant and Animal Based Composites*, K. Kumar and J. P. Davim, Eds. Berlin: De Gruyter, 2021, pp. 25–38.
5. Pinterest, *Pinterest*. [Online]. Available: https://www.pinterest.com/search/pins/?q=plastic%20soil%20composite&rs=typed&term_meta=plastic%7Ctyped&term_meta=soil%7Ctyped&term_meta=composite%7Ctyped (accessed: June 28, 2021).
6. S. Guo, S. Zheng, Y. Hu, J. Hong, X. Wu, and M. Tang, "Embodied energy use in the global construction industry," *Applied Energy*, vol. 256, p. 113838, 2019, doi: 10.1016/j.apenergy.2019.113838.
7. L. Hollaway and P. R. Head, *Advanced Polymer Composites and Polymers in the Civil Infrastructure*. Amsterdam and London: Elsevier Science, 2001.

8. J. Deng and M. M. K. Lee, "Rehabilitation of civil structures using advanced polymer composites," in *Advanced Civil Infrastructure Materials.* V. M. Karbhari, Ed. Sawston, UK: Woodhead Publishing , 2006, pp. 211–234.
9. E. Ferretti and G. Pascale, "Some of the Latest Active Strengthening Techniques for Masonry Buildings: A Critical Analysis," *Materials,* vol. 12, p. 1151, 2019, doi: 10.3390/ma12071151.
10. Designing Buildings Ltd, *Lift Standards: EN 81–20 and EN 81–50.* [Online]. Available: https://www.designingbuildings.co.uk/wiki/Lift_Standards:_EN_81-20_and_EN_81-50 (accessed June 28, 2021).
11. C. V. Amaechi, C. O. Agbomerie, E. O. Orok, and J. Ye, "Economic aspects of fiber reinforced polymer composite recycling," in *Encyclopedia of Renewable and Sustainable Materials.* I. Choudhury and S. Hashmi, Eds. Amsterdam, Netherlands: Elsevier, 2020, pp. 377–397.
12. J. Kim, J. Lee, C. Jo, and C. Kang, "Development of low cost carbon fibers based on chlorinated polyvinyl chloride (CPVC) for automotive applications," *Materials & Design*, vol. 204, no. 6, p. 109682, 2021, doi: 10.1016/j.matdes.2021.109682.
13. H. Shagwira, F. Mwema, T. Mbuya, and A. Adediran, "Dataset on impact strength, flammability test and water absorption test for innovative polymer-quarry dust composite," *Data in Brief,* vol. 29, p. 105384, 2020, doi: 10.1016/j.dib.2020.105384.
14. L. C. Hollaway, "Advanced fibre-reinforced polymer (FRP) composite materials in bridge engineering: materials, properties and applications in bridge enclosures, reinforced and prestressed concrete beams and columns," in *Advanced Fibre-Reinforced Polymer (FRP) Composites for Structural Applications.* J. Bai, Ed. Sawston, UK: Woodhead Publishing, 2013, pp. 582–630.

Processing, Testing, and Failure Modes in Polymer Materials

2

2.1 PROCESSING AND BONDING METHODS OF THERMOPLASTIC COMPOSITES

Polymers have unique and attractive material properties which depend on the long chains in their molecular structures and the type of production process used to manufacture products [1]. The properties of the final product can be predicted based on both the composition (such as cross-linking, molecular size, branching, chemical makeup, and molecular size) and processing (influenced by direction and speed of flow, and orientation of materials' particles) [2]. There exists a wide range of natural and synthetic polymers in the familiar classification of materials such as adhesives, rubbers, fibres, and plastics. These categories of materials should be considered as structural polymers since macroscopic mechanical behaviour forms the basis of their function [3]. The current wide use of polymers has been attributed to the growth in research and development in comparison to traditional structural materials, such as metals.

The areas of composites and polymer blends are rapidly growing [4]. In composite material, a material with a fixed form, for example, fibres or

DOI: 10.1201/9781003231936-2

particles (fillers), is distributed through a polymer matrix. The fibre or filler can either be organic or inorganic, while the polymer matrix can be inorganic. On the other hand, the blends (or alloys) are comprised of two or more polymers that are combined and are different from the composites, as the phase geometry is not predetermined before processing. Polymers find applications in very many fields like engineering, medical, and aerospace.

It may be essential to join different parts during the processing of polymer composites and the manufacturing of products. Joining polymer composites is problematic since conventional joining technologies used in metals and other materials cannot be used directly [5]. In lightweight applications, polymer composites play a very significant role. Achievement of sufficient interface bonding strength is the main challenge in their application. Specific characterization and process monitoring techniques must be developed to optimize inter-facial strength. Various techniques must be studied to determine how the bonding of polymer composites can be carried out to obtain quality bonding strength, without warping and deconsolidation. The widely known bonding technique has been heating and melting these polymers only on the bonding surface and then pressing together the components to form a fusion bond [6].

Design considerations should ideally include constructing structures without joints as they may be sources of weaknesses and weight addition to the structures. However, in the real manufacturing processes, the size of a component is limited by the available resources such as the capacities of the machines. Additional inspection, accessibility, repair, and transport or assembly requirements imply that load-bearing joints are also key components of a structure. This is especially true in the utilization of polymer composites in the construction of large and complex structures. This requires engineers to join different parts with fairly simple geometries limited by the high viscosity of the melt resin and the continuous reinforcement constraints [7].

For instance, some of the challenges experienced in using mechanical fasteners to join composites include [8]:

- The concentration of stress around the holes and cut-outs, exacerbated by the absence of plasticity limiting stress redistribution.
- During drilling operations, localized wear may cause delamination.
- A huge difference between the thermal expansion of composite as compared to fasteners.
- The intrusion of water between composite and fasteners.
- Arcing between fasteners and electrical continuity in the composite.
- Fastened joints may experience galvanic corrosion.
- Fasteners tend to add weight to the structure.
- A lot of time and labour are needed during the drilling of holes.

The thermosetting adhesive bonds are desirable over mechanical fasteners since they are continuously linked and prevent high stresses at each discrete hole of fasteners [9]. Nevertheless, substrate surfaces may be affected by a wide range of contaminants. These include fingerprints, machining oils, carbon release sprays, silicon from release agents, and some of the composite parts which have shifted to the surface (plasticizers, water, calcium stearate, etc.). These must be avoided through surface preparation treatment before bonding is carried out. The treatment can also improve low-energy surfaces wetting, modify the surface chemically (for instance, introducing coupling agents and polar groups), and enhance surface roughness (increasing the bonding surface area and mechanical interlocking capabilities). Extensive preparation of the surface and long cure times intensifies the adhesive bonding work.

2.1.1 Filament Winding

This is an automated process of open moulding that utilizes a rotating mandrel as a mould. The surfaces of the male moulds produce a well-finished inner surface of a product and a laminate surface on the product's external diameter [10]. This process usually leads to high fibre loading, enhancing the manufacture of hollow and cylindrical products such as pressure vessels, tubes, stacks, chemical and rocket motor cases, and fuel storage tanks with relatively high tensile strength. The process usually produces high strength-to-weight ratio composites, offering fibre orientation control and a high degree of uniformity [11, 12]. The filament winding operation can be applied in the construction of highly engineered structures, meeting strict tolerances. Due to automation in filament winding, the factor of minimum labour is required compared to the open moulding process. The process of wet filament winding is displayed in Figure 2.1.

2.1.2 Compression Moulding

This is a type of moulding process where an open, heated, and moulded cavity is filled with the preheated polymer. The mould cavity is then shut with a top plate and compressed such that the polymer material is in full contact with the mould. In this process, a wide range of lengths, thicknesses, and complexity of parts can be produced. These products are normally very strong and attractive for various industrial applications.

The four main steps in the compression moulding process are:

- The production of a metallic tool with high strength, which is machined to have the exact inner dimensions as the intended product/part. The tool is then installed in a press and heated.

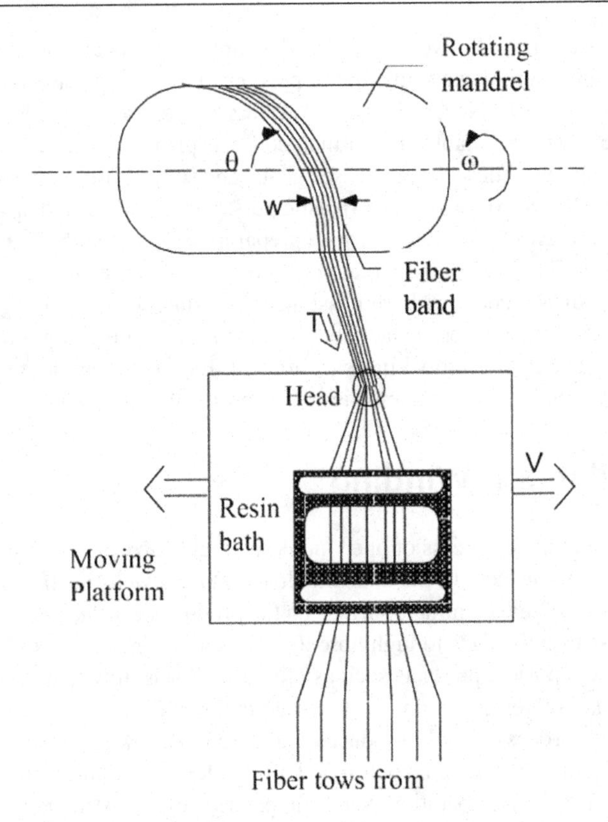

FIGURE 2.1 Schematic diagram of the wet filament winding process [13] (reused with permission from Elsevier Ltd).

- The pre-forming of the composite is needed in the tool/mould shape. The pre-forming of the finished part is a vital step in improving its performance.
- The composite pre-formed is then placed in the preheated mould. Based on the part's thickness and material type used, the material is then compressed under high pressure.
- The pressure is then released, and the produced part is removed from the mould.

2.1.3 Hand Lay-Up Manufacturing

The simplest composite processing technique is the hand lay-up method. This process needs minimal resources and is characterized by very straightforward workflows. During the process, a release gel is first sprayed to prevent the polymer from sticking on the surface of the mould. Twisted or trimmed strand mats to be used as reinforcements are then cut according to the mould size and put on the mould surface. Thereafter, mixing of the liquid thermosetting polymer and a prescribed hardener (curing agent) is thoroughly done in suitable proportions and poured onto the mat surface that has already been put in the mould. The polymer is evenly spread using a brush. The mat's second layer is then put on the polymer surface. With mild pressure, a roller is moved on the mat-polymer to get rid of the excess polymer and any trapped air. The process is repeated until the necessary layers are stacked for every polymer and mat layer. The mould is then opened after curing, and the produced composite component is removed for further processing.

The hand lay-up process is illustrated in Figure 2.2.

2.1.4 Electron Beam Processing of Polymers

Electron beam (EB) processing is an environment-friendly methodology for graft polymerization, cross-linking, and chain scission of polymer materials [15]. A high-energy electron beam accelerator is used in the treatment of polymer composites. Usually, a nitrogen environment is used at suitable temperatures depending on the type of polymer matrix and the reinforcing material. The EB-processed products have proved to perform effectively at high temperatures. They also have good ageing characteristics with only small quantities of chemicals added into the processing system compared to conventional methods [16].

2.1.5 Pultrusion Processing

Pultrusion is a process of continuous moulding using polyester or thermoset-ting matrices reinforced with fibres. Pre-set reinforcement materials, like fibre-glass, are drawn through a bath of resin in which a liquid thermosetting resin is

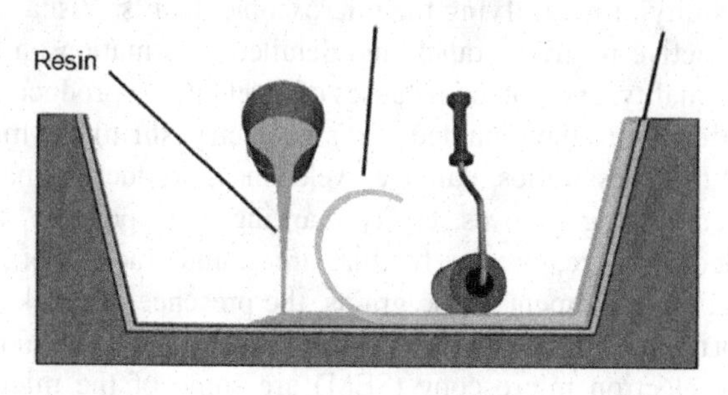

FIGURE 2.2 Hand lay-up process [14] (reused with permission from Elsevier Ltd).

completely impregnated with all materials. The wet-out fibre is developed in the intended geometric form and is fed into a heated steel die. The resin cure is commenced once the wet-fibres are inside the heated steel, and it's done through the accurate control of high temperatures. The solidification of the laminate occurs in the exact cavity pattern of the die, as the pultrusion machine constantly pulls it.

2.2 TESTING POLYMER COMPOSITES USED IN CONSTRUCTION

Polymer composites used for construction purposes are subject to extreme loading conditions and harsh environmental factors. Some of these factors include tensioning, bending, high-energy radiation, weathering (caused by snow, rain, and sunlight), chemical agents, and fire [17]. It is crucial to consider whether environmental and mechanical factors affect the polymer composites at the same time. Environmental stress cracking (ESC) affects polymer composites by causing crack failure [18]. It is quite impossible to predict how polymer composites can behave when affected by all the factors. Therefore, it is important to perform all kinds of tests on a polymer composite to understand its actual behaviour under different environmental conditions. The test preferences should be on the final produced part.

Some of the tests carried out on polymer composites are discussed below.

2.2.1 Visual Inspection and Microstructural Analysis

Optical testing comprises microscopic testing and visual inspection using the naked eye. The optical methods of testing vary on how details can be resolved and the possibility of magnifying the microscopic details. Visual inspections are highly effective in giving quick and detailed information on the design, manufacture, quality, and potential causes of a defect in a product. It is important to pay attention to the moulded part's geometry, shrinkage marks, sharp corners, wall thickness ratios, damage, weld lines, production marking, burr formation, demoulding grooves, cavity marking, gate position, matt/glossy areas, moulded part size, streaks, fracture areas, and cracks. For microstructural analysis, the alignment of the grains, the presence of cracks and voids and any abnormalities in the microstructure are inspected. Optical microscopy and scanning electron microscopy (SEM) are some of the microstructural analyses that can be carried out.

If a polymer composite material is damaged or has failed, it is necessary to evaluate the following [19]:

- Position of failure.
- Type and cause of failure.
- Properties of the part during production and usage.

2.2.2 Physical Tests

The mass is usually a material property of a sample and is responsible for its weight. The weighting of the sample helps in determining its absolute or relative mass for comparisons. An analytical balance is used to accurately determine very small masses and mass differences. A sample mass can be determined for various purposes:

- In determining material density (for a known sample volume).
- In studying the swelling and shrinkage behaviour.
- In determining the ash content.
- In investigating homogeneity of a moulding.
- In determining the volatile content in the sample.
- In studying and optimizing the process of filling injection moulds.

Mass measurement can also be used to determine the sample's moisture absorption. The moisture absorbed is physically associated with hydrogen bridges within the structure of the polymer composite. Water absorption tends to cause the polymer composite to swell, making its volume change. The geometrical dimension of the moulded part and its thermomechanical properties are therefore modified. Additionally, swelling changes the polymer composite's glass transition temperature to a much lower temperature. Therefore, the polymer composite becomes soft and loses rigidity and strength in comparison o drying conditions. With increased moisture content, the material deformaion capacity increases. The actual moisture content for the produced composte dictates the magnitude of geometric and material changes. For a quality ssessment of a polymer composite that has been produced from hygroscopic naterial, its moisture content must be evaluated and defined.

.2.3 Mechanical Tests

Iechanical tests are usually performed to study and ascertain the mechanical roperties of samples. Since products and structures operate under different conditions, mechanical tests are also performed under different testing conditions. These conditions can be dynamic, static, long term, quasi-static, or short term [20]. Mechanical tests involve the deformation of samples to some extent hence they are characterized by strains and stresses. The types of stresses are tensile, compression, bending, and shear stresses. The purpose of carrying out mechanical tests is to ensure the safety of any particular product. The tests are meant to verify that the designs were carried out perfectly, hence encouraging the evolution of technology that will promote good designs.

Some of the mechanical tests that can be carried out on polymer composites include:

- Tensile testing.
- Compressive testing.
- Shear strength testing.
- Hardness testing (Ball, Vickers, Knoop, Durometer, and International Rubber Hardness Degree).
- Impact strength testing.
- Tear resistance testing.
- Dynamic mechanical analysis.
- Abrasion resistance with the Taber Abraser.
- Coefficient of friction of plastic film.
- Flexural strength testing.
- Creep testing.

2.2.4 Chemical Analysis

Polymer composites easily swell when they come in contact with some chemicals. Swelling is the increase in the volume of a product as a result of swelling agent absorption. Swelling causes dimensional changes which make the intermolecular binding forces in the polymer composite weak. This in turn increases the molecular mobility hence reducing the material's stiffness. Additionally, there is a decrease in the glass transition temperature, the mechanical strength, and the hardness of a polymeric material if allowed to swell due to some chemical action on it. The chemical agent may affect a polymer composite and make it unfit for use. Environmental stress cracking (ESC) in the plastics can take place when the material is concurrently exposed to mechanical stresses as a result of either residual stresses in the material, by a chemical agent or external forces [21]. Environmental stress cracking is a typical and crucial polymer composite phenomenon, since, when a material is stressed and in contact with a chemical, it becomes fragile and brittle, even if a mechanical load is applied below the nominal failure limit [22]. The stress cracking resistance of polymeric composite materials should therefore be studied.

Three important test methods used in the determination of ESC of polymers are [23]:

- Ball or pin indentation technique (ISO 22088-4).
- Bending strip technique (ISO 22088-3).
- Tensile creep test (ISO 22088-2).

2.3 FAILURE IN POLYMER COMPOSITES USED IN CONSTRUCTION

Polymer composite failure typically falls within one of the following three classes: design failures, defects in both material and manufacturing, and anomalies in service [24, 25]. The tasks of failure investigation in polymer composites are generally difficult since the failure nature is brittle, and once they fail, many loose fractured pieces are usually present, which may be hard to identify their origin [26, 27]. However, it is important to follow the following procedures before starting.

First, information should be collected on the history and conditions of the failed component. Where possible, a visit to the site of failure should be made to provide details on the positioning of the failed parts before they are collected and inspected. Secondly, protecting fracture surfaces and stabilizing the parts are conducted, which ensure that there is no post-failure damage [28].

Some faces that have fractured, e.g. those fractured through compression, appear inseparable, making it relatively simple to protect the surfaces. However, interlaminar and tensile fractures are more likely to separate in case of failure, and hence vulnerable to post-failure damage [29]. Therefore, protection of fracture surfaces against mechanical damage, e.g. fretting, contamination, and moisture, is necessary.

Failure mode can be ageing, sink marks, fatigue fracture, temperature influences, cracks, deformation, incomplete geometry, and brittle or ductile fractures. Likely hotspots that may experience failures are weld lines, sharp cross-section transitions, edges, bores, radii, notches, corners, gates, insert locations, and ribs [30]. The in-service anomalies may cause a component to fail due to a particular degrading factor that the component was exposed to during its service. Other failure modes include:

- Overload failures.
- Failures caused by poor bonding.
- Failures due to embedded defects.

- Impact-induced failures.
- Failure due to environmental factors.

2.4. SUMMARY

This chapter presents the processing, testing, and failure modes in polymer materials. The wide application of polymer composites in the construction of buildings and other structures necessitates the study of their processing, joining, and failure testing. This is because of the high requirement for safety in the construction industry. The unpredictability of properties of polymer composites requires in-depth research before these materials can be used in producing any product. This chapter therefore has discussed some of the areas of focus in using polymer composites in the construction industry.

REFERENCES

1. E. Oromiehie, A. K. Gain, and B. G. Prusty, "Processing parameter optimisation for automated fibre placement (AFP) manufactured thermoplastic composites," *Composite Structures*, vol. 272, no. 8, p. 114223, 2021, doi: 10.1016/j.compstruct.2021.114223.
2. X. Wang, Z. Guan, X. Liu, Z. Li, G. Han, Q. Meng, and S. Du, "Prediction of the inter-fiber mechanical properties of composites: part I standardization micro-scale modelling method and damage analysis," *Composite Structures*, vol. 271, 4–5, p. 114127, 2021, doi: 10.1016/j.compstruct.2021.114127.
3. A. Pariyar, C. S. Perugu, L. S. Toth, and S. V. Kailas, "Microstructure and mechanical behavior of polymer-derived in-situ ceramic reinforced lightweight aluminum matrix composite," *Journal of Alloys and Compounds*, vol. 880, p. 160430, 2021, doi: 10.1016/j.jallcom.2021.160430.
4. M. Muhammed Shameem, S. M. Sasikanth, R. Annamalai, and R. Ganapathi Raman, "A brief review on polymer nanocomposites and its applications," *Materials Today: Proceedings*, vol. 45, no. 2, pp. 2536–2539, 2021, doi: 10.1016/j.matpr.2020.11.254.
5. A. Yousefpour, "Joining: thermoplastic composites fusion bonding/welding," in *Encyclopedia of Composites*, L. Nicolais, A. Borzacchiello, and S. M. Lee, Eds Hoboken, NJ: Wiley Online Library, 2012.
6. A. Yousefpour, M. Hojjati, and J.-P. Immarigeon, "Fusion bonding/welding of thermoplastic composites," *Journal of Thermoplastic Composite Materials*, vol. 17, no. 4, pp. 303–341, 2004, doi: 10.1177/0892705704045187.

7. T. Joppich, A. Menrath, and F. Henning, "Advanced molds and methods for the fundamental analysis of process induced interface bonding properties of hybrid, thermoplastic composites," *Procedia CIRP*, vol. 66, no. 2, pp. 137–142, 2017, doi: 10.1016/j.procir.2017.03.275.
8. O. Eroğlu, H. J. Langeheinecke, N. Enzinger, and F. Fischer, "Contribution to mechanical fasteners for composite structures – an automated industrial approach," *Materials Today: Proceedings*, vol. 34, no. 3, pp. 326–331, 2021, doi: 10.1016/j.matpr.2020.05.657.
9. C. Ageorges, L. Ye, and M. Hou, "Advances in fusion bonding techniques for joining thermoplastic matrix composites: a review," *Composites Part A: Applied Science and Manufacturing*, vol. 32, no. 6, pp. 839–857, 2001, doi: 10.1016/S1359-835X(00)00166-4.
10. X. Zhang, P. Wang, D. Sun, X. Li, J. An, T. X. Yu, E. Yang, and J. Yang, "Dynamic plastic deformation and failure mechanisms of individual microcapsule and its polymeric composites," *Journal of the Mechanics and Physics of Solids*, vol. 139, p. 103933, 2020, doi: 10.1016/j.jmps.2020.103933.
11. Z. Cui, Q. Liu, Y. Sun, and Q. Li, "On crushing responses of filament winding CFRP/aluminum and GFRP/CFRP/aluminum hybrid structures," *Composites Part B: Engineering*, vol. 200, no. 3, p. 108341, 2020, doi: 10.1016/j.compositesb.2020.108341.
12. Q. Wang, T. Li, B. Wang, C. Liu, Q. Huang, and M. Ren, "Prediction of void growth and fiber volume fraction based on filament winding process mechanics," *Composite Structures*, vol. 246, no. 11, p. 112432, 2020, doi: 10.1016/j.compstruct.2020.112432.
13. L. Zhao, S. C. Mantell, D. Cohen, and R. McPeak, "Finite element modeling of the filament winding process," *Composite Structures*, vol. 52, 3–4, pp. 499–510, 2001, doi: 10.1016/S0263-8223(01)00039-3.
14. M. Raji, H. Abdellaoui, H. Essabir, C.-A. Kakou, R. Bouhfid, and A. e. k. Qaiss, "Prediction of the cyclic durability of woven-hybrid composites," in *Durability and Life Prediction in Biocomposites, Fibre-Reinforced Composites and Hybrid Composites*. M. Jawaid, M. Thariq and N. Saba, Eds. Sawston, UK: Woodhead Publishing, 2019, pp. 27–62.
15. J. G. Drobny, "Electron beam processing of commercial polymers, monomers, and oligomers," in *Ionizing Radiation and Polymers*. J. G. Drobny, Eds. Norwich, United States: William Andrew, 2013, pp. 101–147.
16. N. L. K. Thiher, S. M. Schissel, and J. L. P. Jessop, "Quantitative comparison of photo- and electron-beam polymerizations based on equivalent primary radical concentration," *Radiation Physics and Chemistry*, vol. 172, no. 2, p. 108808, 2020, doi: 10.1016/j.radphyschem.2020.108808.
17. G. Allen, Ed., *Comprehensive Polymer Science: The Synthesis, Characterization, Reactions and Applications of Polymers*. Oxford: Pergamon Press, 1989.
18. Y. Liu, F. P. van der Meer, L. J. Sluys, and L. Ke, "Modeling of dynamic mode I crack growth in glass fiber-reinforced polymer composites: fracture energy and failure mechanism," *Engineering Fracture Mechanics*, vol. 243, p. 107522, 2021, doi: 10.1016/j.engfracmech.2020.107522.
19. Q. Sun, G. Zhou, Z. Meng, M. Jain, and X. Su, "An integrated computational materials engineering framework to analyze the failure behaviors of carbon fiber reinforced polymer composites for lightweight vehicle applications,"

Composites Science and Technology, vol. 202, p. 108560, 2021, doi: 10.1016/j.compscitech.2020.108560.

20. S. Chabira, S. Mohamed, and B. Hadj aissa, "The use of mechanical testing in the study of plastic films degradation," *Revue des Sciences et Sciences de l'Ingénieur*, vol. 03, pp. 28–32, 2013.
21. M. Contino, L. Andena, and M. Rink, "Environmental stress cracking of high-density polyethylene under plane stress conditions," *Engineering Fracture Mechanics*, vol. 241, p. 107422, 2021, doi: 10.1016/j.engfracmech.2020.107422.
22. D. Zhao, Y. Zhou, F. Xing, L. Sui, Z. Ye, and H. Fu, "Bond behavior and failure mechanism of fiber-reinforced polymer bar–engineered cementitious composite interface," *Engineering Structures*, vol. 243, p. 112520, 2021, doi: 10.1016/j.engstruct.2021.112520.
23. ISO 22088-3:2006(en), *Plastics — Determination of Resistance to Environmental Stress Cracking (ESC) — Part 3: Bent Strip Method*. [Online]. Available: https://www.iso.org/obp/ui/ (accessed June 26, 2021).
24. N. Li and Y. D. Zhou, "Failure mechanism analysis of fiber-reinforced polymer composites based on multi-scale fracture plane angles," *Thin-Walled Structures*, vol. 158, p. 107195, 2021, doi: 10.1016/j.tws.2020.107195.
25. E. S. Greenhalgh, Ed., *Failure Analysis and Fractography of Polymer Composites*. Cambridge, UK: Woodhead Publishing and Boca Raton, FL: CRC Press, 2009.
26. E. S. Greenhalgh, *Failure Analysis and Fractography of Polymer Composites*. Cambridge, UK: Woodhead Publishing, 2009. [Online]. Available: http://search.ebscohost.com/login.aspx?direct=true&scope=site&db=nlebk&db=nlabk&AN=689126
27. E. Greenhalgh, "Failure analysis and fractography of polymer composites," in *Failure Analysis and Fractography of Polymer Composites*. E. Greenhalgh, M. J. Hiley, and C. B. Meeks, Eds. Bosa Roca: Taylor & Francis Inc, 2009, pp. 1–595.
28. C. V. Opelt, G. M. Cândido, and M. C. Rezende, "Compressive failure of fiber reinforced polymer composites – a fractographic study of the compression failure modes," *Materials Today Communications*, vol. 15, pp. 218–227, 2018, doi: 10.1016/j.mtcomm.2018.03.012.
29. Z. Ren, L. Liu, Y. Liu, and J. Leng, "Damage and failure in carbon fiber-reinforced epoxy filament-wound shape memory polymer composite tubes under compression loading," *Polymer Testing*, vol. 85, p. 106387, 2020, doi: 10.1016/j.polymertesting.2020.106387.
30. Z. Peng, X. Wang, and Z. Wu, "A bundle-based shear-lag model for tensile failure prediction of unidirectional fiber-reinforced polymer composites," *Materials & Design*, vol. 196, p. 109103, 2020, doi: 10.1016/j.matdes.2020.109103.

Life Cycle Analysis of Plastic

3

3.1 INTRODUCTION

The increasing quantities of waste due to mass consumption has become one of the major societal problems and has been becoming bigger and worse day by day [1–3]. The waste menace can be traced back to the Middle Age period whereby the accumulation of wastes from food would attract a huge number of rats that hosted fleas and enhanced their population growth leading to serious health problems due to plague transmission to humans. This led to the formulation of methods for managing food wastes in big cities to improve hygiene. The main method was to transport the food wastes and faeces outside the city to eliminate the rats and reduce plague transmission but at the same time make agricultural soil fertile [4, 5]. This was the start of waste management systems.

During the ancient civilization, the waste majorly comprised renewable and natural materials like fur, wood, wool, etc., which were used to make materials that could be reused to a point where they could not be repaired. Therefore, the generated wastes during this period were not sufficient to cause major societal problems. However, in the eighteenth century, the industrial revolution was associated with mass production and increased the use of non-renewable sources which had a negative impact. The use of fossil fuels would not only deplete the coal reserves but also cause city pollution, causing health problems. In the twentieth century, artificial materials started to be manufactured, mainly from petroleum, and they were carelessly disposed of. To solve the waste problem, depositing wastes in landfills became the only option [5–10]. The landfill problem has been a concern in the modern world which has caused the formulation of various means of handling wastes.

DOI: 10.1201/9781003231936-3

3.2 THE CONCEPT OF LIFE CYCLE ANALYSIS

Several techniques are used to quantify the environmental impacts to be used in policy implementation [7, 11]. Among them is life cycle analysis (LCA). LCA is a methodology for evaluating the environmental effects and possible impact of a service or a product throughout its life cycle, from cradle to grave [12–14]. This implies taking into consideration all methods taking place from the extraction process of raw materials up until the final waste disposal, including manufacturing, transportation, use, and recycling stages of a product as well. It is generally used in analyzing waste management systems. It can be combined with economic valuation to give a broader perspective to decision-makers by putting all the processes together and the related environmental impacts [13].

There has been an increased interest in the use of LCA over the years which has led to the harmonization of the methodology. In 1991, for the first time, the Society of Environmental Toxicology and Chemistry published guidelines on how to compile the LCA. Other guidelines were then published later. In 1997, International Standards Organization (ISO) published LCA guidelines under ISO 14040 and later ISO 14044 [15]. LCA is used by companies and non-governmental organizations (NGOs) to either identify improvements in the environment or in comparing competitors' products. There are many expectations from LCA, but at the same time, its results are often criticized [16–18].

3.3 PURPOSE OF LCA

3.3.1 Goal

LCA is a tool used to analyze environmental burdens associated with recycled plastics. It aims to give an understanding of the extent to which the production and processing of recycled plastic impacts the environment.

3.3.2 Scope of the Study

The functional unit for a particular study can be set, for instance, as 1 kg o 1 ton of waste plastics at the factory gate (cradle to gate). All the steps fc plastic production must be taken into account. The impact categories in terms of climate change, fossil depletion, acidification potential, eutrophication, and human toxicity must also be assessed. Such categories are chosen because they are widely accepted and understood [19].

3.3.3 Inventory Analysis

In this section, primary data from recycled material producers and secondary data from databases are used. Such data include:

- The amount of waste plastics collected per day.
- The number of trips made by waste collecting trucks per day.
- The amount of fuel consumed by each truck per day.
- The number of trucks available.
- Distance travelled by each truck per day.
- The amount of energy required to process the recycled plastics.

The processing requirements such as electricity, heat production, production of additional substances, and transport can be modelled using software such as Ecoinvent 2.2. After obtaining these set data, openLCA and SimaPro, software such as GaBi, openLCA, and Umberto can be used to conduct life cycle inventory analysis and impact calculations associated with recycling of polypropylene (PP) [20].

3.4 LCA IN WASTE MANAGEMENT

LCA has been used for a long time in making comparisons between different waste treatment alternatives and comparing integrated waste management systems which are complex including the fractions in the waste. Applying LCA in managing wastes has the benefit of helping to have a wide perspective of the analysis and evaluate the whole system, all the processes involved, and the related environmental impacts. By using this technique, a balance of environmental loads can be achieved in separate geographical regions and impact categories such as acidification, type of environment (land, water, air), or different waste management system steps. Various software tools and models have been developed to enable the implementation of LCA and assist in decision making in waste management [20].

For there to be a sustainable waste management system, it must be economically affordable, environmentally efficient, and socially accepted. An economically affordable system means that the costs of managing waste systems are satisfactory to every sector of the community, i.e. individuals, enterprises, industries, and the government, among others. An environmentally efficient system is the one in which there is an environmental burden reduction in managing wastes in terms of the production of harmful emissions to land, water, and air and consumption of resources. A socially accepted system is one in which the system for managing waste meets the local community needs without undermining the priorities and values of the society.

3.5 DETERMINATION OF ENVIRONMENTAL CREDIT

Properties of plastics may be affected by environmental conditions or the working conditions such as loading during service. These properties may not be affected if such conditions are closely monitored. However, it's important to determine the optimum blending ratio for recycling polymers. A comparison between the properties of the virgin and recycled polymer can then be made. Down cycling is defined by the quality factor. This is the ratio of the quality of the recycled polymer (experimentally determined property) versus the quality of the virgin polymer. The quality factor can be used to calculate the environmental credit given by Equation 1 [21]:

$$\text{Environmental Credit} = x \times \text{Recyl} + (1 - x) \times Q \times \text{Virg} \tag{1}$$

Where:

x = the amount of recycled material in the blended mixture
Recyl = environmental load during recycling
$(1 - x)$ = the amount of virgin material in the blended mixture
Q = determined quality factor
Virg = environmental load during the production of virgin material

Applying this formula to LCA of waste management systems to determine the environmental credit for PP-recycled material will help increase the amount of recycled material in the market mixes and come up with a more established waste management system. The room for improvement of waste management systems greatly depends on the average mix of the virgin and recycled material and at the same time the ability of the recycled material to actually find the same applications as that of virgin material with almost the same properties.

The use of conventional means to treat and handle the final disposal of wastes causes pollution and other environmental impacts. Therefore, there is a need for improving measures to reduce and recycle waste materials. There is a need for developing a system for managing wastes that are environmentally effective, socially acceptable, and economically affordable.

Europe has among the best waste management systems in the world. The European Union employs the Waste Directive in managing wastes. In 1975, the first Waste Directive was implemented in Europe [22]. This entailed dealing with wastes once it has already been created. A solution to curb the generation of wastes and depletion of resources could not be established. For this reason, the management of wastes in Europe started focusing on life cycle thinking not only to prevent the generation of wastes but also in the recycling of wastes instead of treating wastes. Waste management policies have also been implemented in the United States by the Environmental Protection Agency of the United States [23]. Its main goal is to prevent wastes generation while promoting recycling and conservation of natural resources. Waste management strategy can follow the waste hierarchy, which was adopted in 1999 from the idea of landfill Directive and the Directive on packaging and packaging wastes [24].

In these Directives, the most important goal is to reduce the amount of waste. If there is waste minimization at the source, there will be less need to handle it at the end of the life of a product. However, there exist laws that only target landfills which set mandatory targets for waste reduction. Improved technologies and a better understanding of pollution have increased the desire to evaluate the environmental, social, and economic benefits and the impacts associated with options of managing wastes, irrespective of where they fall in the waste hierarchy.

Developing countries should adopt the waste management strategies like the European Landfill Directive. The huge amounts of wastes in most developing countries can be turned into a source of energy or resources for industries and therefore wastes should not be seen as a societal problem. The reduced environmental impacts resulting from waste generation should be enhanced in the whole life cycle of a product instead of in the waste phase only [11]. Such countries should encourage the reuse, recycling, and remanufacture of products and materials hence promoting conservation of natural resources. This will prioritize a green economy development in the world.

Using the LCA, more responsible types of legislation can be put in place to reduce wastes. The landfill Directive tends to reduce the amount of biodegradable wastes that cause landfills and bans hazardous wastes from most landfills. Additionally, producers having the responsibility to organize recovery and recycling of wastes is a great step in waste minimization. There must be advantages of recycling waste plastics and not just recycle for the sake of recycling. One of the advantages of recycling waste plastics is the reduction of the demand for crude oil. Additionally, plastic waste recycling reduces emissions and improves energy efficiency as compared to the production of virgin plastics. Energy efficiency depends on several factors which include the grade of the plastic, its condition, and the recycling method. The use of the mechanical recycling method is more efficient than the chemical recycling method [25].

3.6 RECYCLING OF POLYMERS

Mechanical recycling entails the shredding of the plastic wastes into pellets which are then moulded again. This must be accompanied by a careful sorting process to avoid contamination that will degrade the material properties. Mechanical recycling is more preferable as compared to chemical recycling because it produces fewer emissions and has low energy requirements [26]. Processing temperature and shear forces affect the material properties. The viscosity and the impact strength may diminish if high temperatures and shear forces are used. The type of recycling machine also affects the quality of recycled plastic. A machine that has a well-controlled recycling process will produce recycled plastic that has good properties as to those of virgin plastic. Chemical recycling is where the polymer is broken down into its monomers. The processes involve the use of glycolysis, methanolysis, aminolysis, hydrolysis, or hydroglycolysis [27]. This is usually an expensive method as compared to mechanical recycling, and it is accompanied by the application of toxic substances. Additionally, the by-products are liquids and gases which can be harmful. However, the process is suitable for mixed plastic wastes that are difficult to sort out. In thermochemical recycling, the waste plastic is decomposed into a condensed mixture at high temperature in the absence of air, in the presence of hydrogen, or through gasification.

In waste management systems, the development of new technologies has allowed the recycling of waste plastics to be more efficient. It was not possible to apply recycled waste plastics in the packaging of foodstuff due to high contamination levels until 1991 when super clean recycling technology was adopted [28]. However, the existing technological barriers tend to limit the recycling of waste materials in large quantities. Besides, some materials are usually better suited to be recycled than others. An example of a material that can be barred from being recycled in large quantities is plastic. Therefore more research is needed to formulate methods that will curb such limitation of recycling wastes.

A driver of polymer recycling is the slow rate of natural decomposition of thermoplastics products. Under ordinary circumstances, a thermoplastic material is non-degradable because no recognized organism can break down its comparatively big molecules. Complex and costly processes must be used to biologically degrade some thermoplastics. Many scientists indicated that thermoplastic flakes should fulfil certain minimum conditions for effective plastic recycling [29–32]. The key factor influencing the recycling suitability of polymer flakes is the amount and nature of impurities in the flakes. For instance, various techniques have been documented for recycling disposable plastic water bottles and other PET bottles or PET blends with other products [33, 34]. These include virgin resin reprocessing, mixing and stabilization, and solutions-based and chemical reactions recycling [27]. There are three primary considerations to account for when it comes to PET recycling: first, waste collection; second, the process of recycling itself; and third, whether or not there is a recycled end product's market.

3.7 APPLICATIONS OF PHYSICALLY REPROCESSED POLYMER PRODUCTS

The sorted polymer waste is processed by the end-user in physical recycling by pelletizing or granulating, melting or partial melting, and extruding to form the final product. During the process, the base polymer is not modified. The flaked, ground, or pelletized plastics are cleaned to remove impurities before melting and reforming the polymer. Distinct plastics may also experience reforming conditions that are different, e.g. different production temperatures, the use of vacuum stripping, or other processes that may affect contaminant concentrations.

Recycled plastic is much more susceptible to hydrolysis and thermal degradation as compared to virgin plastic [26]. In addition, recycling of plastic results in a reduction in average molecular weight, melt viscosity, and mechanical and thermal characteristics of the material due to the thermomechanical degradation disintegration of the hydrolytic chain during processing [27]. Therefore, processing aids, extra antioxidants, or other additives may need to be applied to the recycled resin in order to create a plastic with the required properties.

During the process, recycled plastics can be added to virgin plastic. Dromiehie and Mamizadeh [35] used separate extrusion techniques to process and modify the mixture of recycled grade (R-PET) and virgin grade PET

(V-PET) with and without a modifier, and then outlined the physical, mechanical, and thermal properties of the modified PET. In the blended samples (R-PET/V-PET), the intrinsic viscosity variation was consistent with the ratio of the two components. It has been shown that when the amount of recycled PET increases, the molecular weight reduces. By adding a compatibilizer to the mixture, these properties were refined.

Blends of polymers have often been known to be an interesting combination for new high-performance polymer material without completely synthesizing new polymers in particular when their properties are combined synergistically. It is furthermore one of the most important routes for polymer recycling because the sorting step is much easier during the recycling process. Most polymers are unfortunately immiscible. Their blends, therefore, demonstrate poor mechanical properties and morphology that is unstable. The use of various types of compatibilizers can solve these problems.

3.8 SUMMARY

In this chapter, an overview of the life cycle analysis of plastics has been presented. The purpose of LCA has clearly been demonstrated and it entails (i) the goal, (ii) the scope of study, and (iii) the inventory analysis. The focus basically being more on the recyclability of plastics to solve the waste issues. It has been demonstrated clearly that the concept of life cycle analysis can be utilized so that the waste management systems can be improved. Improved waste management systems mean plastics can be recycled or reused in making other products. Such products can find their applications in various fields such as construction industries. Once these waste plastics have been utilized in making products such as roofing tiles, the environmental credit can then be calculated.

REFERENCES

1. Z. Bao, W. Lu, and J. Hao, "Tackling the 'last mile' problem in renovation waste management: a case study in China," *Science of the Total Environment*, vol. 790, no. 4, p. 148261, 2021, doi: 10.1016/j.scitotenv.2021.148261.
2. K. Dianati, L. Schäfer, J. Milner, A. Gómez-Sanabria, H. Gitau, J. Hale, H. Langmaack, G. Kiesewetter, K. Muindi, B. Mberu, N. Zimmermann, S. Michie

and M. Davies, "A system dynamics-based scenario analysis of residential solid waste management in Kisumu, Kenya," *Science of the Total Environment*, vol. 777, no. 6, p. 146200, 2021, doi: 10.1016/j.scitotenv.2021.146200.

3. S. Oduro-Kwarteng, R. Addai, and H. M.K. Essandoh, "Healthcare waste characteristics and management in Kumasi, Ghana," *Scientific African*, vol. 12, no. 1, e00784, 2021, doi: 10.1016/j.sciaf.2021.e00784.
4. F. Havlíček, A. Pokorná, and J. Zálešák, "Waste management and attitudes towards cleanliness in Medieval Central Europe," *Journal of Landscape Ecology*, vol. 10, no. 3, pp. 266–287, 2017, doi: 10.1515/jlecol-2017-0005.
5. D. E. Avcı and H. D. Çeliker, "Waste management in ancient times and today from the perspective of teachers: reflections to diaries," *European Journal of Economics and Business Studies*, vol. 1, no. 1, p. 8, 2015, doi: 10.26417/ejes.v1i1.p8-13.
6. G. M. Escandar and A. La Muñoz de Peña, "Multi-way calibration for the quantification of polycyclic aromatic hydrocarbons in samples of environmental impact," *Microchemical Journal*, vol. 164, no. 4, p. 106016, 2021, doi: 10.1016/j.microc.2021.106016.
7. K. Islam, X. Vilaysouk, and S. Murakami, "Integrating remote sensing and life cycle assessment to quantify the environmental impacts of copper-silver-gold mining: a case study from Laos," *Resources, Conservation and Recycling*, vol. 154, p. 104630, 2020, doi: 10.1016/j.resconrec.2019.104630.
8. A. Assuah and A. J. Sinclair, "Solid waste management in western Canadian First Nations," *Waste Management (New York, N.Y.)*, vol. 129, pp. 54–61, 2021, doi: 10.1016/j.wasman.2021.05.007.
9. A. K. Awasthi, V. R. S. Cheela, I. D'Adamo, E. Iacovidou, M. R. Islam, M. Johnson, T. R. Miller, K. Parajuly, A. Parchomenko, L. Radhakrishan, M. Zhao, C. Zhang, and J. Li, "Zero waste approach towards a sustainable waste management," *Resources, Environment and Sustainability*, vol. 3, no. 2, p. 100014, 2021, doi: 10.1016/j.resenv.2021.100014.
10. R. S. Mor, K. S. Sangwan, S. Singh, A. Singh, and M. Kharub, "E-waste management for environmental sustainability: an exploratory study," *Procedia CIRP*, vol. 98, no. 10, pp. 193–198, 2021, doi: 10.1016/j.procir.2021.01.029.
11. E. Bracquené, M. G. Martinez, E. Wagner, F. Wagner, A. Boudewijn, J. Peeters, and J. Duflou, "Quantifying the environmental impact of clustering strategies in waste management: a case study for plastic recycling from large household appliances," *Waste Management (New York, N.Y.)*, vol. 126, pp. 497–507, 2021, doi: 10.1016/j.wasman.2021.03.039.
12. B. K. Sharma and M. K. Chandel, "Life cycle cost analysis of municipal solid waste management scenarios for Mumbai, India," *Waste Management (New York, N.Y.)*, vol. 124, pp. 293–302, 2021, doi: 10.1016/j.wasman.2021.02.002.
13. M. Smith, A. Bevacqua, S. Tembe, and P. Lal, "Life cycle analysis (LCA) of residential ground source heat pump systems: a comparative analysis of energy efficiency in New Jersey," *Sustainable Energy Technologies and Assessments*, vol. 47, no. 3, p. 101364, 2021, doi: 10.1016/j.seta.2021.101364.
14. M. Ilyas, F. M. Kassa, and M. R. Darun, "Life cycle cost analysis of wastewater treatment: a systematic review of literature," *Journal of Cleaner Production*, vol. 310, no. 4, p. 127549, 2021, doi: 10.1016/j.jclepro.2021.127549.
15. ISO, *ISO 14040:2006*. [Online]. Available: https://www.iso.org/standard/37456.html (accessed July 1, 2021).

16. A. Wilfart, A. Gac, Y. Salaün, J. Aubin, and S. Espagnol, "Allocation in the LCA of meat products: is agreement possible?," *Cleaner Environmental Systems*, vol. 2, no. 3, p. 100028, 2021, doi: 10.1016/j.cesys.2021.100028.
17. K. Goulouti, P. Padey, A. Galimshina, G. Habert, and S. Lasvaux, "Uncertainty of building elements' service lives in building LCA & LCC: What matters?," *Building and Environment*, vol. 183, no. 6, p. 106904, 2020, doi: 10.1016/j.buildenv.2020.106904.
18. H. Ismail and M. M. Hanafiah, "An overview of LCA application in WEEE management: current practices, progress and challenges," *Journal of Cleaner Production*, vol. 232, pp. 79–93, 2019, doi: 10.1016/j.jclepro.2019.05.329.
19. T. H. Christensen, A. Damgaard, J. Levis, Y. Zhao, A. Björklund, U. Arena, M. A. Barlaz, V. Starostina, A. Boldrin, T. F. Astrup, and V. Bisinella, "Application of LCA modelling in integrated waste management," *Waste Management (New York, N.Y.)*, vol. 118, pp. 313–322, 2020, doi: 10.1016/j.wasman.2020.08.034.
20. BlueMouse GmbH, *Resellers – Ecoinvent.* [Online]. Available: https://www.ecoinvent.org/partners/resellers/resellers.html (accessed June 26, 2021).
21. A. B. Gala, M. Raugei, and P. Fullana-i-Palmer, "Introducing a new method for calculating the environmental credits of end-of-life material recovery in attributional LCA," *International Journal of Life Cycle Assessment*, vol. 20, no. 5, pp. 645–654, 2015, doi: 10.1007/s11367-015-0861-3.
22. R. Stewart, "Waste management," in *Management, Recycling and Reuse of Waste Composites.* V. Goodship, Ed. Bosa Roca: Taylor & Francis Inc, 2010, pp. 39–61.
23. M. N. Araya, "A review of effective waste management from an EU, national, and local perspective and its influence: the management of biowaste and anaerobic digestion of municipal solid waste," *JEP*, vol. 09, no. 06, pp. 652–670, 2018, doi: 10.4236/jep.2018.96041.
24. S. van Ewijk and J. A. Stegemann, "Limitations of the waste hierarchy for achieving absolute reductions in material throughput," *Journal of Cleaner Production*, vol. 132, no. 2, pp. 122–128, 2016, doi: 10.1016/j.jclepro.2014.11.051.
25. H. Jeswani, C. Krüger, M. Russ, M. Horlacher, F. Antony, S. Hann, and A. Azapagic, "Life cycle environmental impacts of chemical recycling via pyrolysis of mixed plastic waste in comparison with mechanical recycling and energy recovery," *The Science of the Total Environment*, vol. 769, p. 144483, 2021, doi: 10.1016/j.scitotenv.2020.144483.
26. D. Briassoulis, A. Pikasi, and M. Hiskakis, "Recirculation potential of post-consumer/industrial bio-based plastics through mechanical recycling – techno-economic sustainability criteria and indicators," *Polymer Degradation and Stability*, vol. 183, no. 20, p. 109217, 2021, doi: 10.1016/j.polymdegradstab.2020.109217.
27. H.-J. Ho, A. Iizuka, and E. Shibata, "Chemical recycling and use of various types of concrete waste: a review," *Journal of Cleaner Production*, vol. 284, p 124785, 2021, doi: 10.1016/j.jclepro.2020.124785.
28. R. Franz and F. Welle, "Contamination levels in recollected PET bottles from non-food applications and their impact on the safety of recycled PET for food contact," *Molecules (Basel, Switzerland)*, vol. 25, no. 21, p. 4998, 2020, doi: 10.3390/molecules25214998.
29. J. Hopewell, R. Dvorak, and E. Kosior, "Plastics recycling: challenges and opportunities," *Philosophical Transactions of the Royal Society of London. Serie*

B, Biological Sciences, vol. 364, no. 1526, pp. 2115–2126, 2009, doi: 10.1098/rstb.2008.0311.
30. D. Foti, "Use of recycled waste pet bottles fibers for the reinforcement of concrete," *Composite Structures*, vol. 96, pp. 396–404, 2013, doi: 10.1016/j.compstruct.2012.09.019.
31. T. U. Chowdhury, M. Amin Mahi, K. A. Haque, and M. M. Rahman, "A review on the use of polyethylene terephthalate (PET) as aggregates in concrete," *MJS*, vol. 37, no. 2, pp. 118–136, 2018, doi: 10.22452/mjs.vol37no2.4.
32. B. Formisano, S. Göttermann, and C. Bonten, "Recycling of cast polyamide waste on a twin-screw-extruder," *AIP Conference Proceedings,* vol. 1779, no. 1, p. 140002, 2016, doi: 10.1063/1.4965582
33. R. Zhang, X. Ma, X. Shen, Y. Zhai, T. Zhang, C. Ji, and J. Hong, "PET bottles recycling in China: an LCA coupled with LCC case study of blanket production made of waste PET bottles," *Journal of Environmental Management*, vol. 260, p. 110062, 2020, doi: 10.1016/j.jenvman.2019.110062.
34. C. Sealy, "Recycling gives a PET new lease of life," *Materials Today*, vol. 45, pp. 6–7, 2021, doi: 10.1016/j.mattod.2021.03.001.
35. A. Oromiehie and A. Mamizadeh, "Recycling PET beverage bottles and improving properties," *Polymer International*, vol. 53, no. 6, pp. 728–732, 2004, doi: 10.1002/pi.1389.

Plastic and Silica Waste

4

4.1 INTRODUCTION

Plastic pollution has become the most significant environmental issue due to its disposal challenges. The trend is postulated to become worse with the increasing population and demand for plastic consumer goods [1]. In most parts of Africa and Asia, plastic disposal is a huge challenge, and it is common in the streets of cities in these areas to see littered plastic carrier bags and packaging containers [2, 3]. According to National Geographic 2017, growth in plastic pollution is enhanced by the lack of efficient garbage collection systems in these parts of the world (https://www.nationalgeographic.com/). In developed countries, there are better garbage collection and recycling systems; however, there is no country yet to fully solve the problem of plastic pollution in the world [4]. The solution to plastic pollution is multifaceted as it requires both national and international regulatory policies and social and behaviour change among the populations [5, 6]. In some countries, such as Kenya, the use of plastic carrier bags has been banned; however, such countries still grapple with the disposal problem of the other types of plastics especially those used in food packaging [6–9].

Quarrying is used in extracting silica stones used for the construction of buildings and structures [10]. Quarrying is extensively carried out in different parts of Kenya to support the ever-growing construction industry, especially around the satellite and major cities. The activities offer economic support to a significant percentage of households in Kenya and, mostly, the youth have greatly benefited from it [11, 12, 13]. However, quarrying is associated with various (adverse) environmental and social impacts [14, 15]. After quarry extraction, the open pits are left with heaps of waste earth. Additionally, the extraction process contaminates the plants and wildlife and poses a risk to the surrounding communities. Efforts towards land reclamation and reforestation are important in dealing with the effects of quarrying in Kenya, although more

DOI: 10.1201/9781003231936-4

needs to be done. In Chapter 5 of this book, the possibility of utilizing quarry dust as a reinforcement for plastic composites for building application is illustrated as a way of dealing with the pollution effect of quarrying.

4.2 PLASTIC POLLUTION AND RECYCLING

The production of plastic grew from 2.3 million tons to 448 million tons between 1950 and 2015 and that by 2050, the production is predicted to grow to 900 million tons (Figure 4.1) [16, 17]. The upward trend is driven by the growing population and increasing demand for use of plastic for food packaging and carrier bags. The use of plastics is preferred due to their beneficial properties such as lightweight, resistance to bacterial attack, ability to form into different products, and high impact resistance [17]. However, its progressive adoption continues to pose a disposal challenge to the generated waste. This is because most of the plastic products are utilized once and then dumped, which has led to exponential growth in plastic waste over the years [8]. In this section, some of the effects of plastic pollution are highlighted with emphasis on Kenya. Additionally, some of the recycling efforts to convert plastic into useful products are illustrated.

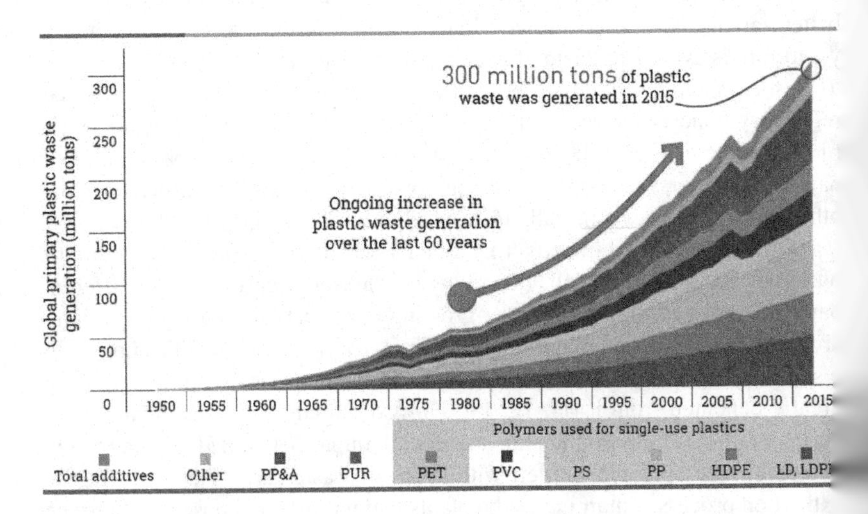

FIGURE 4.1 Global generation of different types of plastics between 1950 and 2015 [17] (reused with permission from Elsevier Ltd).

In Kenya, plastic pollution is highly pronounced in cities and major towns [18]. According to World Bank (www.worldbank.org/en/news) in 2021, for instance, Nairobi produces an estimated 2400 tons of solid waste daily, out of which 20% is plastic waste. It is important to note that of the plastic waste, only 45% is recycled, reused, or transformed in useful forms. This is very low compared to the National Environmental Management Authority's (NEMA) target of 80% reuse and recycling of plastic. Most of the plastic solid waste in Kenya is disposed in dumping sites which are poorly managed in most cities [2, 6]. For example in Nairobi, the plastic waste is usually piled up at the Dandora dumpsite (Figure 4.2) and the recycling and reuse efforts are left to individual Kenyans who collect the plastics and sell them to private enterprises [7, 18]. This approach to recycling and reuse has been ineffective and as such the dump sites continue to grow to huge heaps and landfills.

In agricultural areas of Kenya such as Nyeri County, plastic pollution has several effects as illustrated in Figure 4.3. It forms heaps of solid waste which may have adverse effects on the agricultural land. In fact, they act as sources of microbial and toxic chemical pollution to the soils [19, 20]. In such areas, solid waste disposal such as plastic has several effects on the environment and natural resources – soils and water. Some of these effects on soils include:

i. Solid waste induces soil contamination from its heavy metal and xenobiotic components.
ii. Solid waste reduces a soil's productive capacity by reducing its hydraulic conductivity and percolation capacity. This is due to

IGURE 4.2 Pictures of Dandora dumpsite in Nairobi showing an individual colcting recyclable plastic waste (obtained for free from www.pinterest.com).

FIGURE 4.3 The effect of plastic dumping in agricultural regions of Nyeri County: (a) hazardous plastic pumping around households, (b) plastic dumping along roads and walking paths, (c) plastic dumping in tree plantations, and (d) waste heaps created by plastic dumping (original photo taken by authors).

its formation of continuous biofilms and discontinuous microbial aggregates.

iii. During decomposition, solid waste alters the soil physicochemical characteristics thereby enhancing low pH and soil acidity. Through enhanced solubilization of trace metal elements, solid waste modifies the redox potential of soils [19].

Some of the effects of solid waste on plants include:

i. Solid waste components and particulate matter therein contaminate plants with heavy metal components. These pollutants are transferred to the rest of the food chain considering that plants are primary producers.
ii. Solid waste ash suppresses plant growth and causes retarded/stunted growth in plants [21].

Effects of solid waste dumping on water resources result from the formation of leachate. Leachate is a mineralized and noxious liquid, which results from biochemical and decomposition reactions of disposed solid waste under anaerobic conditions [19]. The liquid is entrained with harmful substances that pollute water resources in several ways:

i. Leachate percolates vertically to introduce heavy metals such as Ni, Pb, Hg, Cd, and Cr and inorganic matter (SO_4, NH_4, CN, Cl, and N) in groundwater resources/aquifers.
ii. Leachate moves laterally to pollute surface water resources such as lakes, creeks, dams, and rivers.

In animals and humans, solid waste and its unscientific disposal results in several impacts:

i. Once ingested, it causes entanglement and eventual death. It also results in low appetite in animals and their eventual starvation to death due to gastrointestinal obstruction [21].
ii. Ingestion of solid waste causes bioaccumulation of toxins in the gut and blood and could lead to liver dysfunction.
iii. In humans, ingestion of solid waste components results in reproductive anomalies, retarded growth, and high vulnerability to respiratory dysfunctions, hormonal disruptions, and genetic mutations that result in cancers [20].

Overall, solid waste disposal results in deterioration of aesthetic beauty and the presence of foul smells that make environs unhealthy and uncomfortable for animals and human habitats.

In most parts of the country, plastic and solid waste recycling is mostly undertaken by individuals, groups, or non-governmental organizations. As shown in Figures 4.4 and 4.5, in Nakuru (Kenya) various groups of individuals including women and youth benefit a lot from picking plastic waste and transforming the plastic waste and other solid waste from the dumpsite into useful products such as shopping bags. As further illustrated by the presence of makeshift houses in Figure 4.5, the dumpsite is a source of livelihood for a considerable number of families in Nakuru. This is typically a demonstration of how recycling plastic waste can support the youth economically across the country and the rest of Africa [22].

Due to lack of proper disposal, the plastic debris ends up in lakes and oceans. According to a report by Thevenon et al. (2014), over 8 million tons of plastic collect as debris in lakes and oceans and constitute up to 80% of

FIGURE 4.4 Picture of Nakuru dumpsite. The women are recycling plastic waste into beautiful shopping bags. The recycling role has been left to individual Kenyans, non-governmental, and private organizations (obtained for free from www.pinterest.com).

FIGURE 4.5 Makeshift houses of the communities surviving on the sale of waste plastic to private recyclers from Nakuru dumpsite (obtained for free from www.pinterest.com).

FIGURE 4.6 Debris along the marine environment (image obtained under open access from [23]).

all marine solid debris [23]. These debris are dangerous to aquatic life since marine animals can ingest these debris causing injuries and death [24, 25]. Most of these plastics originate from human activities around the waters bodies and others are dumped by beach visitors and industrial activities around the oceans (Figures 4.6 and 4.7). In rivers, plastic waste cause clogging and degrades the quality of the water for human/animal use [24].

The solution to the plastic menace is recycling and reusing and reduction of dumping. In Figure 4.8, several applications of waste plastics in the sustainable/green construction industry are shown. As shown, the waste plastic containers and bags can be recycled and transformed into home décor, interior design, and furniture for both indoor and outdoor uses (Figure 4.8a). It can also be recycled with silica/cement/natural fibres to produce composite material for brick and block manufacturing (Figure 4.8b). In Nigeria, waste plastic bottles have been utilized in walling of buildings, (Figure 4.8c) and finally, they can also be used n the fabrication of coloured/aesthetically appealing tiles. Table 4.1 provides information regarding applications of typical virgin and waste plastic.

4.3 QUARRY EXTRACTION POLLUTION

s earlier mentioned, quarry stones are extensively used as building materials n the Kenyan construction sector [11, 12]. However, their extraction leads to nvironmental pollution in several ways.

FIGURE 4.7 Plastic pollution around the ocean in Thailand. Tragically, a pilot whale died in Thailand after having 17 pounds of plastic removed from its stomach (obtained for free from www.pinterest.com).

i. There is a generation of dust in the surrounding communities (Figure 4.9).
ii. The pits are usually left open and they can hold stagnated water which becomes breeding sites for mosquitoes as shown in Figure 4.10. The pits are also risky to the communities around them as an individual can fall and get serious injuries or even die.
iii. The excavated pits are also a threat to the existing plants around them since they expose the roots of surrounding huge trees and plants (Figure 4.11).
iv. The mining activities are a threat to the water bodies such as rivers around them; the huge heaps of quarry dust may leach into the rivers and other water bodies hence covering them which makes them dry up (Figure 4.12).

FIGURE 4.8 Applications of recycled plastic waste: (a) home decorations and tools, (b) building bricks, (c) plastic bottle wall, and (d) plastic tiles (obtained from www.pinterest.com).

The quarry dust can be reutilized with polymeric materials to produce composites which can find applications in the construction industry. A study [26] by the authors of this book reported on the data characterization of polymer-silica based composite for building application, and it was demonstrated that the composite can find application as a roofing tile. The authors have reviewed the properties of related composites [27] and it is no doubt that silica-based particles can reinforce waste plastic for construction applications.

4.4 STATE-OF-THE-ART REVIEW OF PLASTIC WASTE UTILIZATION IN CONSTRUCTION

he increasing impact of plastic waste on the environment has attracted the tention of researchers to recycle or transform them into useful products. In

TABLE 4.1 Typical Plastics, Recycling, and Applications (reused from Almeshal et al. [17] with permission from Elsevier Ltd)

PLASTIC NAME	*ELABORATION*	*SEVERAL APPLICATIONS FOR RECYCLED PLASTIC*	*SEVERAL APPLICATIONS FOR VIRGIN PLASTIC*
PET: Polyethylene Terephthalate	May be used as a fibre, clear tough plastic	Detergent bottles, soft drink bottles, fleece jackets	Mineral water and soft drink bottles
UPVC: Un-plasticized Polyvinyl Chloride	May be clear, hard rigid plastic	Detergent bottles, plumbing pipe fittings, and tiles	Juice bottles and clear cordial, plumbing pipes fittings, and blister packs
HDPE: High-Density Polyethylene	Very common plastic, usually white or coloured	Detergent bottles, compost bins, crates, mobile rubbish bins	Freezer bags, crinkly shopping bags, milk and cream bottles
LDPE: Low-Density Polyethylene	Flexible, soft plastic	Industry, packaging and plant nurseries, film for builders and bags	Garbage bags, garbage bins, lids of ice-cream containers, and black plastic sheet
PPVC: Plasticized Polyvinyl Chloride	Elastic-plastic, flexible and clear	Industrial flooring and hose inner core	Shoe soles, garden hose, tubing, and blood bags
PS: Polystyrene	May be clear, glassy, rigid, brittle plastic	Office accessories, coat hangers, spools and rulers, and clothes pegs	Plastic cutlery, yoghurt containers, and imitation crystal "glassware"
PP: Polypropylene	Many uses, flexible and hard plastic	Kerbside recycling crates and compost bins	Potato crisp bags drinking straws, ice-cream containers
EPS: Expanded Polystyrene	Lightweight, foamed, thermal insulation and energy-absorbing	Non-recycled	Meat trays and packaging, takeaway food containers, and hot drink cups

(Continued)

TABLE 4.1 (CONTINUED) Typical Plastics, Recycling, and Applications (reused from Almeshal et al. [17] with permission from Elsevier Ltd)

PLASTIC NAME	*ELABORATION*	*SEVERAL APPLICATIONS FOR RECYCLED PLASTIC*	*SEVERAL APPLICATIONS FOR VIRGIN PLASTIC*
ABS: Acrylonitrile Butadiene Styrene	Rigid, opaque, glossy tough, good low-temperature properties	Furniture such as chairs and tables, luggage cases and containers	Telephone handsets, rigid luggage, electroplated parts, radiator grills
PC: Polycarbonate	Transparency, strong, thermal stability, stiff, hard, tough, very rigid	Data storage including, CD and DVD and construction materials such as dome lights and sound walls	Automotive, glazing, electronics, business machine, optical media and medical equipment, lighting
HIPS: Polystyrene (High Impact)	Hard, rigid, translucent, high impact strength	Countertop point of purchase displays and indoor signs	Yoghurt pots, refrigerator linings, vending cups, bathroom cabinets, toilet seats, and tanks
PA: Polyamides (Nylon)	Very tough materials with good thermal and chemical resistance	Hair combs, machine screws, gaskets, and other low- to medium-stress components	Fishing line, carpets, food packaging, offering toughness, and low gas permeability
TPE: Thermoplastic Elastomers	Flexible, clear, elastic, wear-resistant, impermeable	Damping elements, grip surfaces, design elements, and back-lit switches and surfaces	Soles and heels for sports shoes, hammerheads, seals, gaskets, and skateboard wheels
EP: Epoxies	Rigid, clear, very tough, chemical resistant, good adhesion properties	Non-recycled	Adhesives, coatings, encapsulation, electrical components, and aerospace applications

FIGURE 4.9 A picture shows the effect of dust to the surrounding community due to quarrying activities. It can be seen that there are households and schools around the quarrying areas and therefore exposing them to the unwarranted dust (photo taken by authors at Chaka quarry site).

this respect, several publications on the recycling of plastic waste into polymer-based composites for various applications are available in the literature [28]. For instance, Lopez et al. [29] developed wood-plastic composites from waste thermoplastics (PTE, HDPE, PP, and calcium carbonate) and *Cedrela odorate L* (sawdust) through the extrusion of different contents of the raw materials. The study revealed that the quantity of the ratio of the sawdust and thermoplastics determined the physical–mechanical characteristics of the composite.

Keskisaari and Karki [30] undertook a cost analysis between using waste and virgin plastic as raw materials for wood-plastic composite. It was reported that using waste plastic reduces the manufacturing cost of wood-plastic composite. This report affirms the need to recycle plastic waste for low-cost housing. In a related study, Hyvärinen et al. [31] investigated the effect of incorporating waste construction materials into wood-plastic composites. It was determined that although the use of waste materials in the fabrication of wood-plastic composite exhibited inferior properties, their Charpy impact strength improved significantly meaning that the composites can find structural applications in the industry sector. In another study, Singh et al. [32] investigated the effect of waste plastic on the properties of banana-plastic composites. It was reported that using waste plastic led to a reduction in tensile strength and an improvement in the tensile modulus. Binhussain and El-Tonsy [33] reported on the fabrication of plastic-wood composite from plastics and palm leaves for outdoor

FIGURE 4.10 Photos showing deep and open pits after quarry excavations at Chaka, Nyeri County, quarry sites (photo taken by authors).

applications in place of natural wood. It was reported that waste plastic-palm leaves composites can replace natural wood in some outdoor applications.

Akbar and Liew [34] in their work titled "assessing recycling potential of carbon fiber reinforced plastic waste in the production of eco-efficient cement-based materials" investigated the potential of utilizing waste carbon-reinforced plastic into the fabrication of cement-plastic composites. In their analyses, they reported that utilizing waste carbon-reinforced plastic into cement-plastic composites contributed to the reduction in carbon dioxide emissions and hence lowering global warming.

The influence of silica-containing agro-wastes in the reinforcement of polymer composites was reported by Rizal et al. in 2020 [35]. In specific, the influence of filler weight ratios and particle sizes of the silica-containing agro-waste (oil palm boiler ash in this case) was investigated, and it was determined that both mechanical and thermal performance of the polymer composites depend on the filler quantities and particle sizes. The use of plastic waste in combination with silica- or/and cement-related materials for construction

FIGURE 4.11 Showing deep and open pit after excavation of quarry stones. Such pits are a threat to the existing trees (photo taken by authors).

waste was emphasized as a scheme to fight environmental pollution in modern society by a critical review presented by Awoyera and Adesina [36]. In this review, recycled plastics were identified to find applications in the following areas of the construction sector:

- A binder in cementitious composites for various applications.
- Base and subbase for road constructions.
- Incorporation into asphalt used in the construction of pavements.
- As a replacement of wooden structures in which the plastic matrix is mixed with sawdust and other fibres.
- Wood-plastic composites can be used as door panels.
- Expanded polystyrene (EPS) can be used as an insulator in buildings.
- Walls and bricks for buildings and related structures.

The review has further highlighted some of the limitations of waste plastic used in the construction sector, some of which include:

- Most of the waste plastics are highly contaminated from various plastics and other chemicals during collection from the dumpsites.

FIGURE 4.12 A photo taken at Chaka quarry site on dust heap. It can be seen that a stream of water next to the quarry mine has been fully covered by the quarry dust (photo taken by authors).

- There is a huge variance in the chemical composition of different plastics, and as such during recycling, a lot of time is consumed in sorting.
- Plastics are lightweight materials and have low density. As such, in cases where toughness is required, their applications may be limited.
- There are no standards currently available on the utilization of plastic materials in the construction industry. This poses a challenge in the design and implementation of constructions utilizing plastic waste.

There are so many other efforts in developing waste-plastic composite materials for the construction sector with the following focus:

i. The influence of the waste plastic in comparison with the virgin plastic in the polymeric composite.
ii. The potential replacement of conventional construction materials with waste polymeric composites.

There is a continuous need to explore further development of composite materials to support the growing construction industry.

4.5 SUMMARY

In this chapter, an overview of plastic and quarry extraction pollution has been briefly presented with a focus on the Kenyan situation. It is stated that the management of plastic waste has been a problem for a long time. As such, the dumping of single-use plastic materials has continued to grow exponentially since 1950. The quantity of plastic waste dumped in landfills/dumpsites has been postulated to grow further in the future. In Kenya, quarry mining is an important economic activity, due to the adoption of quarry stones in construction by society. However, the extraction activities cause environmental pollution in several ways as discussed in this chapter. The emphasis of this chapter is on the need to embrace recycling, reuse, and transformation of waste plastic and silica-based materials such as quarry dust to produce quality materials for the construction industry. As such, a state-of-the-art review of some of the recent works on the waste polymer composite materials for the construction sector has been highlighted. It was discovered that the focus of the research efforts are twofold: (i) comparison between virgin and waste plastic for composite applications, and (ii) replacement of the conventional building materials with the waste polymeric composites. The utilization of waste plastic and other readily available reinforcement materials (such as silica and waste cement from collapsed building debris) can be a significant contributor to the low-cost housing in Kenya and other developing countries. Such focus, additionally, shall contribute to environmental protection, sustainability, and green building/construction technologies. In Chapter 5, a case study on the preparation and testing of plastic-quarry dust composite from the original data of the authors has been presented as a material for sustainable and green construction in developing countries.

REFERENCES

1. K. S. Khoo, L. Y. Ho, H. R. Lim, H. Y. Leong, and K. W. Chew, "Plastic waste associated with the COVID-19 pandemic: crisis or opportunity?," *Journal of Hazardous Materials*, vol. 417, p. 126108, 2021, doi: 10.1016/j.jhazmat.2021.126108.

2. O. O. Ayeleru, S. Dlova, O. J. Akinribide, F. Ntuli, W. K. Kupolati, P. F. Marina, A. Blencowe, and P. A. Olubambi, "Challenges of plastic waste generation and management in sub-Saharan Africa: a review," *Waste Manag.*, vol. 110, pp. 24–42, Jun. 2020, doi: 10.1016/j.wasman.2020.04.017.
3. R. K. Henry, Z. Yongsheng, and D. Jun, "Municipal solid waste management challenges in developing countries – Kenyan case study," *Waste Management*, vol. 26, no. 1, pp. 92–100, 2006, doi: 10.1016/j.wasman.2005.03.007.
4. M. L. Van Rensburg, S. L. Nkomo, and T. Dube, "The 'plastic waste era'; social perceptions towards single-use plastic consumption and impacts on the marine environment in Durban, South Africa," *Applied Geography*, vol. 114, p. 102132, 2020, doi: 10.1016/j.apgeog.2019.102132.
5. B. Sharma, Y. Goswami, S. Sharma, and S. Shekhar, "Inherent roadmap of conversion of plastic waste into energy and its life cycle assessment: a frontrunner compendium," *Renewable and Sustainable Energy Reviews*, vol. 146, p. 111070, 2021, doi: 10.1016/j.rser.2021.111070.
6. J. Njeru, "The urban political ecology of plastic bag waste problem in Nairobi, Kenya," *Geoforum*, vol. 37, no. 6, pp. 1046–1058, 2006, doi: 10.1016/j.geoforum.2006.03.003.
7. L. Oyake-Ombis, B. J. M. van Vliet, and A. P. J. Mol, "Managing plastic waste in East Africa: niche innovations in plastic production and solid waste," *Habitat International*, vol. 48, pp. 188–197, 2015, doi: 10.1016/j.habitatint.2015.03.019.
8. J. Carlos Bezerra, T. R. Walker, C. A. Clayton, and I. Adam, "Single-use plastic bag policies in the Southern African development community," *Environmental Challenges*, vol. 3, p. 100029, 2021, doi: 10.1016/j.envc.2021.100029.
9. L. C. Mascarenhas, B. Ness, M. Oloko, and F. O. Awour, "Multi-criteria analysis of municipal solid waste treatment technologies to support decision-making in Kisumu, Kenya," *Environmental Challenges*, vol. 4, p. 100189, 2021, doi: 10.1016/j.envc.2021.100189.
10. L. T. Bui, P. H. Nguyen, and D. C. M. Nguyen, "Model for assessing health damage from air pollution in quarrying area – case study at Tan Uyen quarry, Ho Chi Minh megapolis, Vietnam," *Heliyon*, vol. 6, no. 9, p. e05045, 2020, doi: 10.1016/j.heliyon.2020.e05045.
11. F. Apollo, A. Ndinya, M. Ogada, and B. Rop, "Feasibility and acceptability of environmental management strategies among artisan miners in Taita Taveta County, Kenya," *Journal of Sustainable Mining*, vol. 16, no. 4, pp. 189–195, 2017, doi: 10.1016/j.jsm.2017.12.003.
12. D. M. Franks, C. Ngonze, L. Pakoun, and D. Hailu, "Voices of artisanal and small-scale mining, visions of the future: report from the International Conference on Artisanal and Small-scale Mining and Quarrying," *Extractive Industries and Society*, vol. 7, no. 2, pp. 505–511, 2020, doi: 10.1016/j.exis.2020.01.011.
13. O. A. K'Akumu, B. Jones, and A. Blyth, "The market environment for artisanal dimension stone in Nairobi, Kenya," *Habitat International*, vol. 34, no. 1, pp. 96–104, Jan. 2010, doi: 10.1016/j.habitatint.2009.07.003.
14. M. A. Mwakumanya, M. Maghenda, and H. Juma, "Socio-economic and environmental impact of mining on women in Kasigau mining zone in Taita Taveta County," *Journal of Sustainable Mining*, vol. 15, no. 4, pp. 197–204, 2016, doi: 10.1016/j.jsm.2017.04.001.

15. A. Kinyondo and C. Huggins, "State-led efforts to reduce environmental impacts of artisanal and small-scale mining in Tanzania: implications for fulfilment of the sustainable development goals," *Environmental Science & Policy*, vol. 120, no. March, pp. 157–164, 2021, doi: 10.1016/j.envsci.2021.02.017.
16. N. H. Zulkernain, P. Gani, N. Chuck Chuan, and T. Uvarajan, "Utilisation of plastic waste as aggregate in construction materials: a review," *Construction and Building Materials*, vol. 296, p. 123669, 2021, doi: 10.1016/j.conbuildmat.2021.123669.
17. I. Almeshal, B. A. Tayeh, R. Alyousef, H. Alabduljabbar, A. Mustafa Mohamed, and A. Alaskar, "Use of recycled plastic as fine aggregate in cementitious composites: a review," *Construction and Building Materials*, vol. 253, p. 119146, 2020, doi: 10.1016/j.conbuildmat.2020.119146.
18. K. Muindi, T. Egondi, E. Kimani-Murage, J. Rocklov, and N. Ng, "'We are used to this': a qualitative assessment of the perceptions of and attitudes towards air pollution amongst slum residents in Nairobi," *BMC Public Health*, vol. 14, no. 1, pp. 1–9, 2014, doi: 10.1186/1471-2458-14-226.
19. J. Nyika, E. Onyari, M. O. Dinka, and S. Mishra, "A review on methods of assessing pollution levels from landfills in South Africa: a review on methods of assessing pollution levels from landfills in South Africa Joan Mwihaki Nyika and Ednah Kwamboka Onyari * Megersa Olumana Dinka Shivani Bhardwaj Mishra," *International Journal of Environment and Waste Management*, vol. 1, April, 2021 (in press).
20. J. M. Nyika, "The plastic waste menace and approaches to its management through biodegradation," in *Handbook of Research on Waste Diversion and Minimization Technologies for the Industrial Sector*, A. K. Rathoure, Ed. Hershey: IGI Global, 2021, pp. 218–235.
21. J. M. Nyika and E. K. Onyari, "A review on solid waste management using the DPSIR framework in a Southern Africa case study," *Handbook of Research on Waste Diversion and Minimization Technologies for the Industrial Sector*, A. K. Rathoure, Ed. Hershey: IGI Global, pp. 13–34, 2020, doi: 10.4018/978-1-7998-4921-6.ch002.
22. M. Nasrollahi, A. Beynaghi, F. M. Mohamady, and M. Mozafari, "Plastic packaging, recycling, and sustainable development," in *Responsible Consumption and Production, Encyclopedia of the UN Sustainable Development Goals*, W. Leal Filho, A. Azul, L. Brandli, P. Özuyar, and T. Wall, Eds. Cham, London: Springer, 2020, pp. 1–9, https://doi.org/10.1007/978-3-319-71062-4_110-1.
23. F. Thevenon, C. Carroll, and J. Sousa, *Plastic Debris in the Ocean: The Characterization of Marine Plastics and Their Environmental Impacts, Situation Analysis Report*. Gland, Switzerland: IUCN. https://portals.iucn.org/library/node/44966. 2015.
24. F. Ronkay, B. Molnar, D. Gere, and T. Czigany, "Plastic waste from marine environment: demonstration of possible routes for recycling by different manufacturing technologies," *Waste Management*, vol. 119, pp. 101–110, 2021, doi: 10.1016/j.wasman.2020.09.029.
25. E. O. Okuku, L. Kiteresi, G. Owato, K. Otieno, J. Omire, M. M. Kombo, C. Mwalugha, M. Mbuche, B. Gwada, V. Wanjeri, A. Nelson, P. Chepkemboi, Q. Achieng, and J. Ndwiga, "Temporal trends of marine litter in a tropical recreational

beach: a case study of Mkomani beach, Kenya," *Marine Pollution Bulletin*, vol. 167, no. March, p. 112273, 2021, doi: 10.1016/j.marpolbul.2021.112273.

26. H. Shagwira, F. Mwema, T. Mbuya, and A. Adediran, "Dataset on impact strength, flammability test and water absorption test for innovative polymer-quarry dust composite," *Data in Brief*, vol. 29, p. 105384, 2020, doi: 10.1016/j.dib.2020.105384.
27. H. Shagwira, F. M. Mwema, and T. O. Mbuya, "Lightweight polymer–nanoparticle-based composites," in *Nanomaterials and Nanocomposites*, First edition. B. S. Babu and K. Kumar, Eds., Boca Raton, FL : CRC Press, 2021, pp. 31–50.
28. H. Shagwira and F. M. Mwema, "Advances in animal/plant–plastic composites: preparation, characterization and applications," in *Plant and Animal Based Composites.* , K. Kumar and J. P. Davim, Eds. Berlin: De Gruyter, 2021, pp. 25–38.
29. Y. Martinez Lopez, J. B. Paes, D. Gustave, F. G. Gonçalves, F. C. Méndez, and A. C. Theodoro Nantet, "Production of wood-plastic composites using cedrela odorata sawdust waste and recycled thermoplastics mixture from post-consumer products - a sustainable approach for cleaner production in Cuba," *Journal of Cleaner Production*, vol. 244, p. 118723, 2020, doi: 10.1016/j.jclepro.2019.118723.
30. A. Keskisaari and T. Kärki, "The use of waste materials in wood-plastic composites and their impact on the profitability of the product," *Resources, Conservation & Recycling*, vol. 134, no. May 2017, pp. 257–261, 2018, doi: 10.1016/j.resconrec.2018.03.023.
31. M. Hyvärinen, M. Ronkanen, and T. Kärki, "The effect of the use of construction and demolition waste on the mechanical and moisture properties of a wood-plastic composite," *Composite Structures*, vol. 210, no. October 2018, pp. 321–326, 2019, doi: 10.1016/j.compstruct.2018.11.063.
32. B. Kumar Singh, U. Kumar Komal, Y. Singh, S. Singh Banwait, and I. Singh, "Development of banana fiber reinforced composites from plastic waste," *Materials Today: Proceedings*, vol. 44, pp. 2194–2198, 2021, doi: 10.1016/j.matpr.2020.12.352.
33. M. A. Binhussain and M. M. El-Tonsy, "Palm leave and plastic waste wood composite for out-door structures," *Construction and Building Materials*, vol. 47, pp. 1431–1435, 2013, doi: 10.1016/j.conbuildmat.2013.06.031.
34. A. Akbar and K. M. Liew, "Assessing recycling potential of carbon fiber reinforced plastic waste in production of eco-efficient cement-based materials," *Journal of Cleaner Production*, vol. 274, p. 123001, 2020, doi: 10.1016/j.jclepro.2020.123001.
35. S. Rizal, H. M. Fizree, M. S. Hossain, Ikramullah, D. A. Gopakumar, E. C. W. Ni, and H. P. S. A. Khalil, "The role of silica-containing agro-industrial waste as reinforcement on physicochemical and thermal properties of polymer composites," *Heliyon*, vol. 6, no. 3, p. e03550, 2020, doi: 10.1016/j.heliyon.2020.e03550.
36. P. O. Awoyera and A. Adesina, "Plastic wastes to construction products: status, limitations and future perspective," *Case Studies in Construction Materials*, vol. 12, p. e00330, 2020, doi: 10.1016/j.cscm.2020.e00330.

Preparation and Characterization of Silicon-Based Plastic Composite

5

5.1 INTRODUCTION

This is an experimental study in which samples of plastic–quarry dust were fabricated through moulding at various amounts/weights of quarry dust. The composite is presented as a case study for polymer-silica composites for potential applications in the construction industry. Characterizations of the various samples were undertaken for microstructure, mechanical, and chemical properties to evaluate the effect of weight of the quarry dust into the polymer-based composites. In this chapter, details of materials and experimental procedures for both sample fabrication and characterizations are presented.

5.2 MATERIALS

The quarry dust used in this study was obtained from Chaka quarry site, location (−0.337083, 36.997401) in Nyeri County, Kenya. The quarry dust was chosen for this research due to its availability and pollution effect at quarry stone mines. The quarry dust was used as obtained from the site, that is, there was no further grinding or sieving. The plastic materials used were virgin high-density polyethene (HDPE) and polypropylene (PP) pellets obtained from TH Wildau

DOI: 10.1201/9781003231936-5

polymer laboratory. The reason for the choice of these plastic materials is due to their attractive mechanical properties and are the most widely used plastics, thus providing an opportunity for recycling [1]. Additionally, there is no much difference in mechanical properties between virgin and recycled PP and HDPE [2, 3].

5.3 SAMPLE PREPARATION

The plastic–quarry samples were fabricated in a polymer laboratory at TH Wildau. The laboratory is sufficiently equipped with facilities for polymers and composites manufacturing and testing. The plastic–quarry dust composite samples for HDPE and PP were prepared in 5wt.%, 20wt.%, 40wt.%, 60wt.%, and 80wt.% composition of quarry dust. The choice of these ratios was justified in twofold reason: (i) where it was found that for lower weight concentrations (less than 20wt.%) of nano-silica in HDPE matrix, 15wt.% produced a composite with the optimal mechanical properties [4]. It was, therefore, necessary to investigate the behaviour of polymer-silica composite at higher concentrations of silica. (ii) It was observed that when a very high concentration of quarry dust was used, large torque was required for effective mixing during fabrication of the composite. The total weight of the masses of the mixture (quarry dust + thermoplastic pellets) was restricted to 300 g because of the limiting capacity of the mixing machine. Additionally, the volume of the components was ensured to be sufficient enough for the cases of high percentage composition of quarry dust so that homogenous mixing could take place.

The mixing of the composite constituents was undertaken using a Brabender Plasti-corder (Model: 625249130, Germany). The mixing equipment has two counter-rotating mixing rotors and three heating walls to enhance the effective/complete melting of the polymer materials. It contains an interface software that is used to monitor the temperature in three heating elements. Additionally, it contains a provision in its control panel for varying and adjusting the speed of the rotors during the mixing process. Throughout the experiment, the three heating elements were kept at a constant temperature of 180°C for HDPE and 200°C for PP. These temperatures were chosen based on their thermal characteristics reported in the literature [5]. For instance heating HDPE beyond 180°C resulted in the formation of char and fumes hence the destruction of the samples. The rotor speed was set at 8.9 rpm (as the safest speed recommended by the experienced operators) and the mixing process was undertaken for 10 minutes since it was observed that homogenization of the composite took place within this time. No abrasion on the machine parts by the quarry dust was observed.

After mixing, the hot composites were collected and taken to a pre-heated hot press (Polystat 400 S, Germany) to form 200 mm × 200 mm × 4 mm sheets according to ISO 293 standard. The hot-pressing machine contains two hydraulically controlled plates. Additionally, heating and cooling chambers are also provided for moulding purposes. The chamber contains a 600 mm × 600 mm area of operations with several sizes of moulds, and it has a maximum pressing force of 450 kN.

The maximum operating temperature of the hot press is 300°C. The temperature was maintained at 180°C, a hydraulic pressure of 20 bars was applied on the material with each sample pressed for 5 minutes before being cooled to 30°C. A silicon paper was used to cover the compressed samples so that they could not stick on the mould compartment. Twelve samples of plastic–quarry dust composite were successfully produced in accordance with each test standards and sliced into different sizes for testing.

5.4 CHARACTERIZATIONS

5.4.1 Charpy Impact Test

The Charpy impact test was conducted according to [6], using a Charpy impact tester (Zwick PSW 4J, Germany). This test is used for determining the quantity of energy absorbed by the material as it fractures. With such a standard high strain rate (400 s^{-1}) testing machine, it is easier to establish the quantity of energy absorbed by a given material during fracture and hence the toughness of the material can be determined, and the brittle-ductile transition can be studied. The samples were cut according to ISO 2818 to avoid introducing defects and points of weaknesses in the sample.

After obtaining the energy reading from the Charpy test scale, the unnotched samples' Charpy impact strength a_{CU} (kJ/m^2) was computed for each sample using Equation 1 [7]. For each sample, ten (10) tests were conducted for statistical accuracy.

$$a_{CU} = \frac{E_C \times 10^3}{h \times b} \tag{1}$$

Where:

E_C = the corrected stored energy in Joules
h = the thickness of the test specimen (mm)
b = the width of the tests specimen (mm)

5.4.2 Optical Microscopy

An optical microscopy, Keyence Optical Microscopy (VK-X1000, Germany), was used to study the morphology of samples for pure PP, pure HDPE, PP + 80wt.% dust, HDPE + 80wt.% dust, and pure dust. The samples were selected since there was no change in the chemical composition of the composite; hence it was suitable to investigate at no mix and at the highest concentration of quarry dust. The fractured samples used in the Charpy impact test were taken and sliced into small pieces for viewing under the microscope. A coherent laser beam is used to illuminate the surface of the specimen to be magnified by the lenses with a magnification range of between 10X and 100X.

5.4.3 Water Absorption Test

Water absorption test was conducted according to ASTM D 570-98 [8]. The composite's water absorption test was used for the determination of the quantity of water absorbed. The sizes of the samples were 60 mm × 60 mm × 1 mm. One set of samples were immersed in water under atmospheric pressure and another set of samples immersed in distilled water under 6 bars. No drying was done prior to immersion, and the initial conditions of the samples were taken to be those of room temperature condition. Roofing tiles are normally used under atmospheric pressure but sometimes there might be a need of using these composites under high pressure, e.g. in underwater applications to mention a few. An airtight stainless steel container was used since it does not easily undergo corrosion. To undertake a high-pressure water absorption test, compressed air was supplied to the container to provide a pressure of 6 bars to the water.

The specimens were immersed in distilled water at 23°C for 24 hours, 1 week, and after every two weeks, until saturation was reached (a change of less than 5 mg). Before weighing was done, the samples were wiped using a tissue to remove water on their surface. The absorption of water is expressed as the percentage increase in weight for each specimen as in Equation 2 [8].

$$\%\,\text{Increase in weight} = \frac{\text{wet weight}, m_2 - \text{conditioned weight}, m_1}{\text{conditioned weight}, m_1} \times 100 \quad (2)$$

The precision of the weighing balance was 0.0001 g as per the standard.

5.4.4 Flammability Test/Limiting Oxygen Index

A flammability test was conducted according to ASTMD 2863 [9] to accurately determine the relative flammability of the composites using the Dynisco

Limiting Oxygen Index Chamber (Model LOI 14273, USA). Dynisco Limiting Oxygen Index Chamber is 50.8 cm high by 30.5 cm wide by 31.5 cm deep and contains a stand, a chimney gas dispersion chamber with glass bead bed, twin gas pressure gauges (0–100 psi), and a rigid sample holder. To safely determine the relative flammability of the samples, it is necessary to measure the minimum oxygen concentration which can support the combustion of the samples. The samples were burnt inside the chamber while precisely regulating the amount of oxygen and nitrogen inside the chamber.

The test was conducted at 21°C with three sets of gases velocity used: normal velocity (4.0 cm^3/sec), low velocity (3.2 cm^3/sec), and high velocity (4.8 cm^3/sec). The normal velocity was first used at 21% oxygen, before proceeding to low velocity then high velocity. The control of the gases was made in small increment and decrement with a waiting time of at least 3 minutes for each step until the flame went off or at least 5 cm of the sample was consumed. The readings of the flow metre were then recorded and were used to calculate the oxygen index according to Equation 3 [9].

$$\mathrm{LOI} = \frac{100 \times \mathrm{O_2\,flow\,rate}\left[\mathrm{cm^3/min}\right]}{\mathrm{O_2\,flow\,rate}\left[\mathrm{cm^3/min}\right] + \mathrm{N_2\,flow\,rate}\left[\mathrm{cm^3/min}\right]} \tag{3}$$

5.4.5 Thermal Analysis

Thermal analysis of the plastic–quarry dust composites was performed according to DIN EN ISO 11357-1:2016(E) [11]. A heat-flux differential scanning calorimetry (DSC 204 Phoenix, Netzsch, Germany) was used. Closed crucibles made up of aluminium were used and three holes were made on their covers for ventilation purposes to avoid pressure changes during measurement and permit the exchange of gases with the surrounding atmosphere. The measured masses of the crucible and the sample are as shown in Table 5.1.

Pure nitrogen gas at a flow rate of 50 ml/min was used to purge the instrument while performing the measurement. Small cuboidal shaped pieces measuring approximately 2 mm × 2 mm × 1 mm were cut from the interior of the composites and were used as the samples. While loading the instrument, the temperature was kept at 30°C above room temperature to prevent condensation of moisture on or inside the crucible. The sample was placed in the crucible carefully to maintain good thermal contact between the sample and crucible. Good thermal contact was also enhanced between the DSC holder and the crucible. The heating range was programmed from –50°C to 300°C at a heating rate of 10°C/min.

TABLE 5.1 Weight Measurements of DSC Crucibles and Samples Prepared in a Different Composition of Quarry Dust

SAMPLE NUMBER	*MATERIAL*	*WEIGHT OF CRUCIBLE (MG)*	*WEIGHT OF THE SAMPLE (MG)*
4343	PP	41.3	8.4
4344	PP + 5% dust	40.9	10.9
4345	PP + 20% dust	41	14
4346	PP + 40% dust	41	15.6
4347	PP + 60% dust	41.6	18.5
4348	PP + 80% dust	41.4	16.1
4349	HDPE + 5% dust	41.3	10.9
4350	HDPE + 20% dust	40.9	10.5
4351	HDPE + 40% dust	40.9	8.6
4352	HDPE + 60% dust	40.9	10.7
4353	HDPE + 80% dust	41.4	19.0
4354	HDPE	41	11.3

5.4.6 Scanning Electronic Microscopy

A scanning electron microscopy (SEM) (model JEOL JSM-6010 LV, TH Wildau, Germany) was used for the analysis and observation of the bonding of the interaction between the quarry dust and plastic interface and obtained the microscopic structural images of the composite. The fractured samples obtained from the impact test were collected and SEM was used at room temperature to observe their cross-section to investigate the plastic and quarry dust interface bonding. The SEM samples were prepared by using a 45 nm thick silver layer sputter coating by applying a 50-mA current at a pressure of 10–12 bars. A high vacuum sputter coater machine (LEICA EM SCD 500, Germany) was used during this process.

5.4.7 Shore D Hardness Test

Shore D Hardness test was conducted as per ASTM D2240 standard [12] and a Shore D Durometer Model PCE-DX-DS was used. It is necessary to determine the relative hardness of the quarry dust–plastic composite as a quality control measure due to its applicability in construction. In this test, the penetration made by the indentor of the test instrument was measured. The samples of thickness 10 mm were placed on a hard and flat surface. While making sure the instrument was perpendicular to the surface of the sample, the indento was pressed into the sample at different points each for one second and the readings were recorded.

5.4.8 Fourier Transformed Infrared Spectroscopy

Infrared spectroscopy is an optical spectroscopy process used to identify the material structure of an unknown polymer sample and to determine the composition of compounds [13]. Infrared spectroscopy is an important method of investigating the composition of the organic compounds, used in the identification of the test sample, and determining its purity. Fourier transform infrared (FTIR) spectroscopy analyses were performed on pure PP, pure HDPE, PP + 80wt.% dust, and HDPE + 80wt.% to determine the interaction between the matrix material and the quarry dust. The change in wt.% composition did not have a significant effect on the chemical structure of the samples hence it was not necessary to study all the samples. An FTIR Spectrometer 640-IR FT-IR, Germany, was used. Due to the same chemical structure within the composite sample, a small piece was cut from each sample and placed in the sample compartment for analysis.

5.5 CHARACTERIZATION RESULTS

The results obtained from various tests were used to evaluate the suitability of using the produced composite for the manufacturing of roofing tiles. The results are presented in the form of tables and figures. Due to the adoption of the methodologies used in other research, comparisons of the results obtained in this research have been made to results obtained in previously conducted researches. This is meant to give a clear understanding of various behaviours of the composite.

5.5.1 Scanning Electron Microscopy (SEM) and Optical Microscopy Analysis

5.5.1.1 *Pure Quarry Dust*

The optical microscopy was used to investigate the optical properties and size of quarry dust used as shown in Figure 5.1a. The results show that the

FIGURE 5.1 Quarry dust images viewed under (a) optical microscope and (b) scanning electron microscope (SEM). (The length of the scale bars shown on the SEM images are 20 µm and 10 µm.)

quarry dust consists of different types of components with different colours, suspected to be oxides. The particle sizes were determined to be between 10 and 20 µm when viewed using light microscopy. This range of particle size denotes an insignificant change in particle sizes, hence, a uniform distribution of particle sizes. This justifies why no further sieving was carried out. Quarry dust particles can be classified as macro-size, micro-size, or nano-size, with the size classification having different impacts on the final produced composite. The particle size measurement was carried out by forming a circle using three sharp edges of several particles. The diameter of the circle formed the average particle size. With the help of SEM, it was possible to observe much smaller particles under 1 µm and the shape of the dust particles was observed to have sharp edges as shown in Figure 5.1b. Through Energy Dispersive X-ray spectroscopy (EDX), the elemental composition of the particles as shown in Figure 5.2, and Table 5.2 indicates that 20% of the composition is made up of silicon and 62% is made up of oxygen; hence a large percentage composed

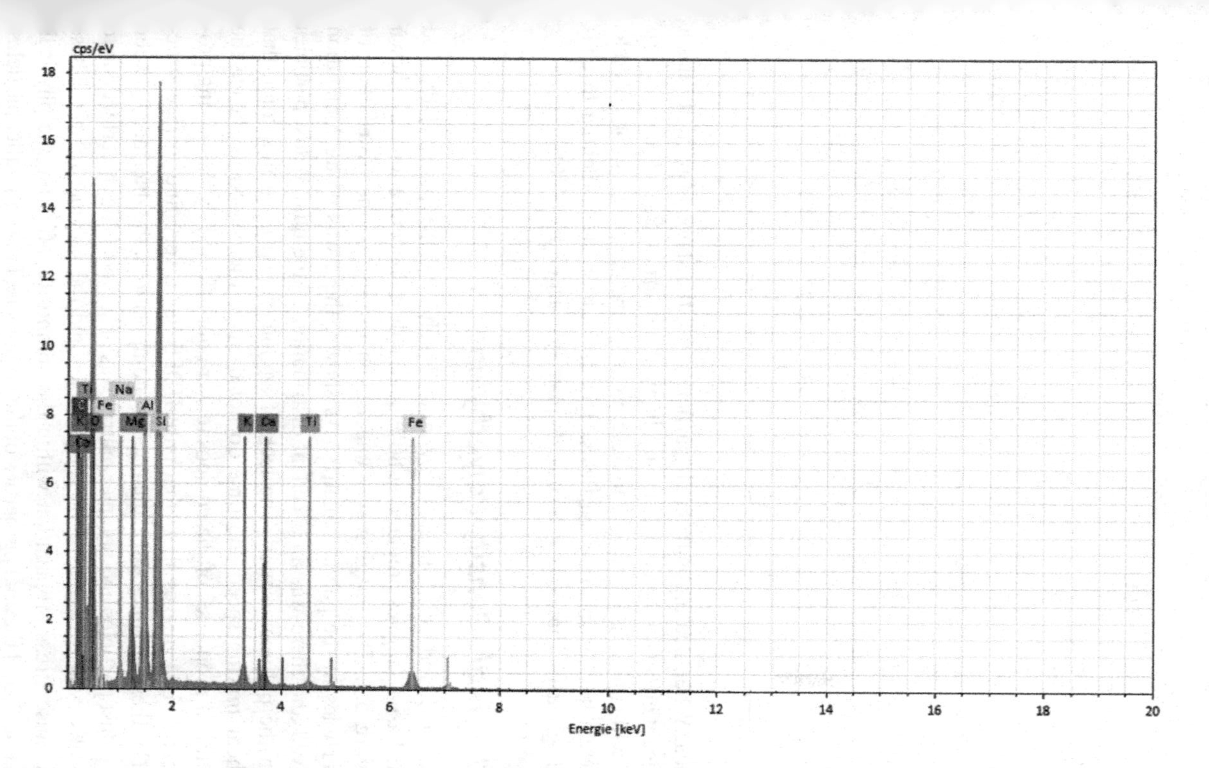

FIGURE 5.2 Energy Dispersive X-ray spectroscopy (EDX) taken on the areas of the images shown in Figure 5.1b showing elemental distribution in quarry dust.

TABLE 5.2 Elemental Distribution in Quarry Dust Sample Determined by Energy Dispersive X-ray Spectroscopy (EDX)

ELEMENT	*ORD. Z.*	*NET*	*MASS %*	*NORM. MASS*	*ATOM*	*ABS ERROR [%] (1 SIGMA)*	*REL. ERROR [%] (1 SIGMA)*
Oxygen	8	45842	61.75	52.44	60.43	7.35	11.91
Silicon	14	78168	20.95	17.79	11.68	0.92	4.41
Carbon	6	2819	12.5	10.62	16.3	2.22	17.73
Aluminium	13	31623	10.44	8.87	6.06	0.53	5.08
Iron	26	4494	3.76	3.2	1.06	0.14	3.8
Magnesium	12	8352	3.33	2.83	2.15	0.22	6.49
Potassium	19	4100	1.55	1.32	0.62	0.08	5.17
Calcium	20	3212	1.41	1.2	0.55	0.08	5.35
Sodium	11	2279	1.38	1.17	0.94	0.13	9.14
Titanium	22	1183	0.68	0.58	0.22	0.05	7.95
		SUM	117.76	100	100		

of silica. When dealing with small particle sizes, the shape of particles has a significant effect on the composite mechanical properties. Particles with sharp edges will cause the composite to have higher principal stresses compared to composite having spherical particles. The dislocation strengthening mechanism is greatly enhanced by the use of spherical particles [14–16].

5.5.1.2 High-Density Polyethylene

The optical and SEM microscopy micrographs for the pure HDPE and HDPE-based samples prepared at 80% quarry dust concentrations are shown in Figure 5.3. The imaging was undertaken to observe the distribution of the particles in the polymer matrix. As shown, there was no agglomeration observed and there was an excellent distribution of the quarry dust particles within the polymer matrix. The lack of agglomeration is an indication that the shear mixing speed and time do not damage the composite; hence the mechanical properties of the composite were not affected by the mixing process. The SEM images show that pure HDPE has a very fine surface structure. For these samples (80% quarry dust composition), the interface between quarry dust particles and polymer matrix was observed to be imperfect due to pores observed between the bigger particles and the polymer matrix. These pores cause stress propagation within the composite, affecting the physical mechanical, and thermal properties of the composite. For the smaller particles a better bonding at the interface was observed. Smaller particles show good interface bonding because of a bigger active surface area. Therefore, for bette

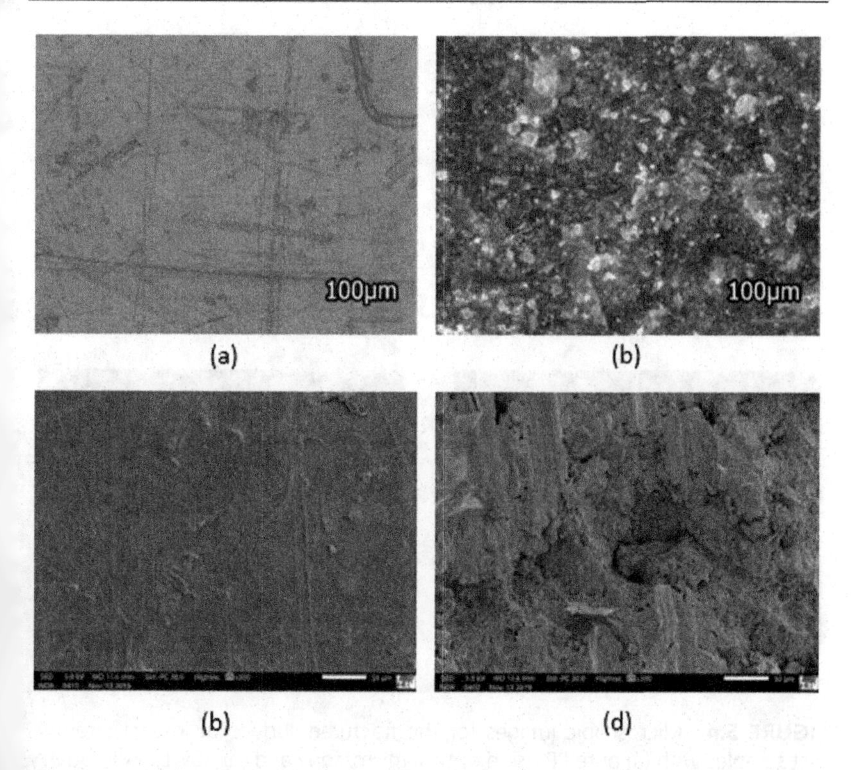

FIGURE 5.3 Microscopic images for the fractured surface of impact strength test samples with (a) pure HDPE using optical microscopy and (b) 80wt.% quarry dust–HDPE composite using optical microscopy (c) pure HDPE using SEM and (d) 80wt.% quarry dust–HDPE composite using SEM.

compatibility between the quarry dust and the HDPE matrix, it is necessary to mill the dust to finer particles, using standard sieving techniques to get the desired particle size. These SEM observations correlate with the impact strength results described later.

5.5.1.3 *Polypropylene*

The microstructural observations of pure PP samples and PP – 80wt.% of uarry dust composites are represented in Figure 5.4. As shown, good distri-ution and particle sizes were observed, which means that the mixing process oes not destroy the composite through agglomeration due to inappropriate hear mixing speed and time. The SEM analysis indicates that pure PP has a ery fine structure of the surface. The interface between quarry dust particles nd polymer matrix was observed not to be good for the 80wt.% dust since

FIGURE 5.4 Microscopic images for the fractured surface of impact strength test samples with (a) pure PP using optical microscopy and (b) 80wt.% PP–quarry dust composite using optical microscopy (c) pure PP using SEM and (d) 80wt.% PP–quarry dust composite using SEM.

there are voids observed between the larger particles and the polymer matrix. As known from literature, cracks and voids negatively affect the mechanical properties of a composite [17]. A stronger bonding was observed at the interface between the smaller quarry dust particles and the polymer matrix. As suggested earlier under the HDPE discussion, it is essential to mill the quarry dust in order to obtain fine particles using standard sieving methods for better mechanical properties of the composite.

5.5.2 Charpy Impact Strength Test on Quarry Dust–Plastic Composite

The Charpy impact test values $\left(a_{UN}\right)$ were obtained for each of the samples and the average values plotted. Figure 5.5 shows the impact strength for PP–quarr

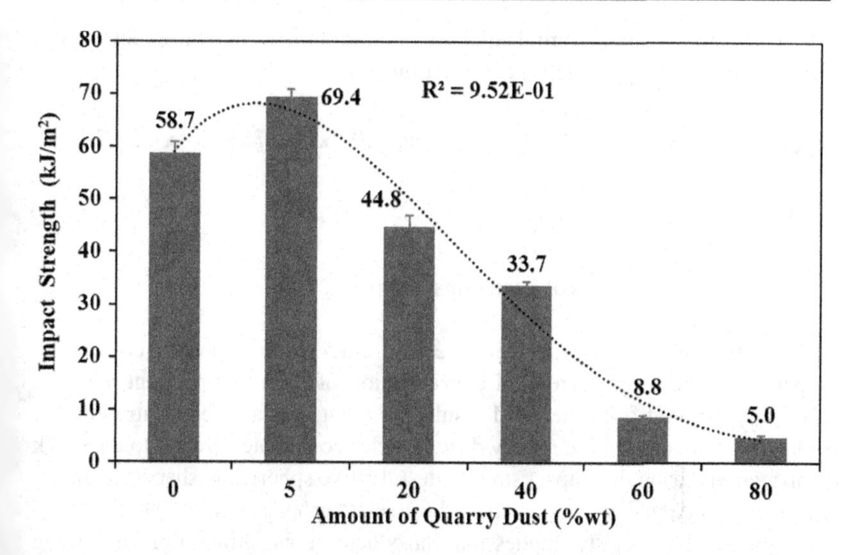

FIGURE 5.5 Impact strength for the PP–quarry dust composite samples prepared at different concentrations of the quarry dust obtained from [10] under open access Creative Commons CC-BY.

dust composite samples prepared at 0wt.%, 5wt.%, 20wt.%, 40wt.%, 60wt.%, and 80wt.% quarry dust concentrations. As shown, the highest impact strength (69.4 kJ/m^2) is observed on samples prepared with 5wt.% concentration of quarry dust, while the lowest impact strength is obtained at 80wt.% composition of quarry dust (5 kJ/m^2). It is further observed that an improvement in the Charpy impact strength value occurs only with the addition of 5wt.% concentration of quarry dust, with the rest of the cases exhibiting lower strengths as compared to the samples without the quarry dust. By increasing the amount of quarry dust beyond the 5wt.% concentration, the impact strength decreases, and the shape of the trend line becomes a polynomial of order 4. The impact strengths obtained for the other concentrations are 44.8 kJ/m^2, 33.7 kJ/m^2, and 8.8 kJ/m^2 for 20wt.%, 40wt.%, and 60wt.% quarry dust compositions, respectively.

The decrease in impact strength beyond 5wt.% quarry dust concentration observed in these results could be attributed to low interfacial adhesion, and therefore, poor transfer of stresses between the polymer and quarry dust phases [18]. During the impact tests, it was observed that between 40wt.% and 80wt.% of quarry dust concentrations, the mode of fracture shifts from ductile to brittle. Similar behaviour has been reported in the literature [19, 20]. Using Equation 4, the relationship between the concentrations and the impact strengths can be well demonstrated. The regression coefficient R^2 =

0.95 shows that the polynomial equation of order 4 fits well to the experimental data; hence it can be used in the predictions.

$$\mathbf{a_{UN}} = -3.11\times10^{-6}\mathbf{x}^4 + 7.92\times10^{-4}\mathbf{x}^3 - 5.86\times10^{-2}\mathbf{x}^2 + 5.38\times10^{-1}\mathbf{x} + 58.7\ldots \quad (4)$$

Where:

$a_{UN} \equiv$ Un-notched impact strength
x ≡ the concentration of quarry dust in % wt

It has been reported in the literature that crack propagation in composites depends on the shape, size, and concentration of the reinforcement particles [21–24]. As noted from the SEM results for 80wt.% quarry concentration, the shape (morphology) of the quarry dust particles contributes greatly to the crack propagation during the impact strength test. Unlike spherically shaped particles, sharped-edged particles tend to allow the crack to propagate faster into the material [25, 26]. It generally implies that introducing concentration of reinforcing particles (such as quarry dust in this case) enhances the rate of crack propagation and material failure under impact loads. Similar results have been reported in published works for glass fibre-reinforced polyester composites [27], glass bead-filled epoxies composite [28] and epoxy–carbon composite, etc. [28–31].

Figure 5.6 shows the impact strengths for HDPE–quarry dust composite samples prepared at 0wt.%, 5wt.%, 20wt.%, 40wt.%, 60wt.%, and 80wt.% concentrations of quarry dust. Similarly, the addition of a 5% weight concentration of quarry dust exhibits a positive effect on the HDPE samples. Further increase beyond this concentration decreases the impact strength of the composite. This inverse proportion relationship is observed in the cases of 20wt.% quarry dust at 24.3 kJ/m^2, 40wt.% quarry dust at 9 kJ/m^2, 60wt.% quarry dust at 5.3 kJ/m^2, and 80wt.% quarry dust at 4.1 kJ/m^2. Similarly, this may be due to the low interfacial adhesion and thus poor stress transfer between the two phases as earlier explained. Additionally, the change of the fracture type from a ductile to a brittle mode from 40wt.% to 80wt.% quarry dust concentration may be due to the increase in the number of the sharp-edged quarry dust particles as observed in Figure 5.1, which tend to fasten the crack propagation unlike spherically shaped particles which can act as crack terminators [19–22] This can also be explained by the fact that the reduction of polymeric materia causes a weak interfacial adhesion, and hence quickens the crack propagation The variation of the impact strength with the concentration of quarry dust ca be demonstrated using Equation 5. The regression coefficient $R^2 = 0.952$ show predictions can be made using the polynomial equation of order 4.

$$\mathbf{a_{UN}} = -8.88\times10^{-6}\mathbf{x}^4 + 1.55\times10^{-3}\mathbf{x}^3 - 8.13\times10^{-2}\mathbf{x}^2 + 8.03\times10^{-1}\mathbf{x} + 30.2\ldots \quad (5$$

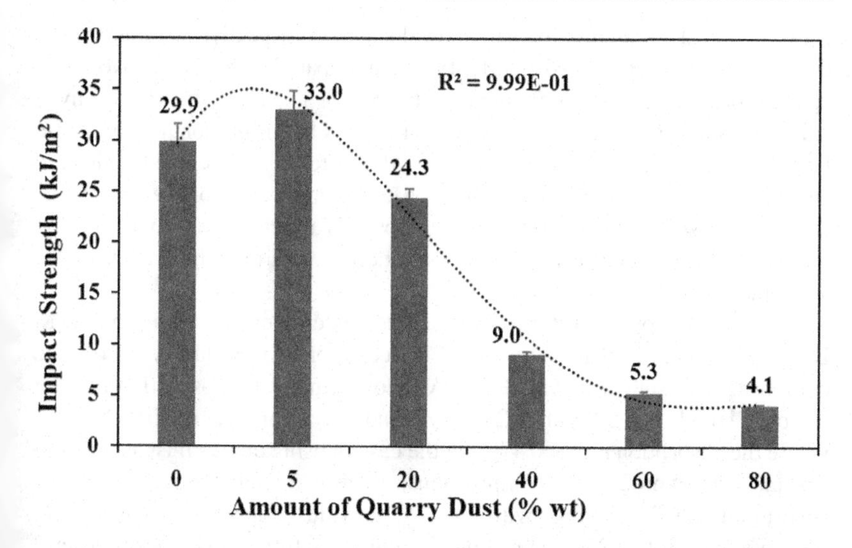

FIGURE 5.6 Impact strength for prepared HDPE–quarry dust composite samples prepared at varying concentrations of quarry dust obtained from [10] under open access Creative Commons CC-BY.

Comparing the results of the impact strength of PP and HDPE composites, it can be seen that (i) the 5wt.% quantity of quarry dust has a positive influence on the impact strength for both composites and (ii) the PP–quarry dust-based composites have much higher impact strength as compared to the HDPE–quarry dust composites and could be attractive in applications requiring higher impact strengths such as floor and roofing tiles. For samples prepared at higher concentrations of quarry dust, the impact strength can be improved by the addition of impact and interfacial modifiers such as graphene oxide (GO), acrylic, low molecular weight hydroxyl-terminated natural rubber (HTNR), and TiO_2 just to mention but a few [32–34].

5.5.3 Water Absorption Test

Many polymeric materials reversibly absorb water from the environment. The moisture absorbed becomes physically attached via hydrogen bridges onto the polymer structure [35]. The absorption of moisture is characterized by polymer polarity. Hydroscopic is the name given to polymeric materials that tend to absorb moisture. The prepared composites are generally hydrophilic and therefore it is necessary to evaluate their behaviour when exposed to water at different pressures (in this case the tests were undertaken at atmospheric pressure and at a pressure of 6 bars to evaluate their respective responses). The objective is to lower the hydrophilicity of the prepared samples for favourable applications such as in buildings and other fields. The mechanical and physical performances of a material can be influenced by dimensional stability (in length, width, thickness, etc.). Therefore, the water absorption test is desired since it can be used to predict the shelf life and application of the composite. The water absorption capability of a composite is affected by the type of polymer used, the exposure time, the type of additive, and the temperature of the surrounding.

The results for the water absorption test undertaken according to ASTM D 570-98 standard [8] for a period of 7 weeks for HDPE–quarry dust-based composites are shown in Figure 5.7. As shown, the water absorption rate for the pure HDPE is 0.28% after 24 hours, and saturation occurs after 3 weeks where the absorption rate is 0.47%. In the case of using quarry dust as an additive (at 5wt.% and 20wt.% compositions), there is a slight decrease in water absorption rate. The sample with 5wt.% quarry dust composition has a water absorption rate of 0.133% after 7 weeks, but the saturation is reached after 3 weeks at 0.105%. In the case of using 20wt.% composition of quarry dust, it is observed that the saturation is at 0.36% after 5 weeks, which is still lower than that of pure HDPE. This shows quarry dust has a positive effect on the polymer matrix as it reduces the rate at which the water is absorbed. With increasing the concentration of quarry dust, the rate of water absorption increases greatly

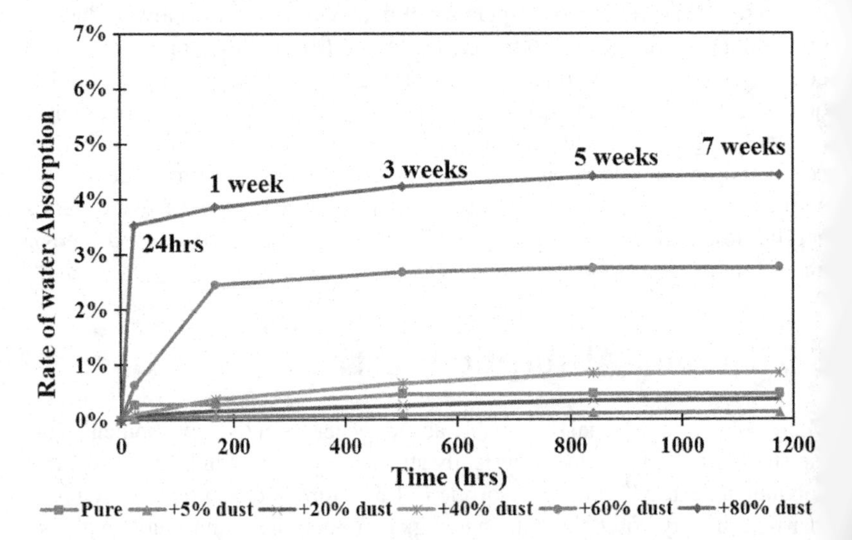

FIGURE 5.7 Water absorption rate at atmospheric pressure for HDPE–quarr dust composite samples prepared at different concentrations of quarry dus obtained from [10] under open access Creative Commons CC-BY.

between 60wt.% and 80wt.% composition of quarry dust. For 60wt.% composition of quarry dust, the saturation is reached after 3 weeks at 2.75%, while for 80wt.% composition of quarry dust, the saturation is reached after 3 weeks at 4.2%. The high water absorption rate at this weight concentrations (60% and 80%) can be attributed to poor adhesion between the interface of the polymeric matrix and the quarry dust and an increased porosity due to the increased amount of quarry dust as it will be shown later by the microstructure results and as reported in published literature [36]. As per the general observation of the trends/shapes of the curves in Figure 5.7, the mass of the absorbed water increases with the square root of time of exposure of the samples to water, obeying Fick's Law, until saturation is achieved [37, 38]. As time increases, the interfacial bonding weakens, increasing the size of the spaces in the composite occupied by water. Water absorption in composites is undesirable since it increases the chances of mechanical failures due to the build-up of internal stresses in the composite making them less durable. The water absorption enhances swelling of the composite, which dislodges the arrangement of the structure of the composites resulting in stress build-up and dimensional changes [39–43]. The dimensional changes may cause fitting problems in the applications such as construction just to mention but a few.

The water absorption rates for PP–quarry dust samples are shown in Figure 5.8. As shown, the water absorption rate saturation for pure PP is

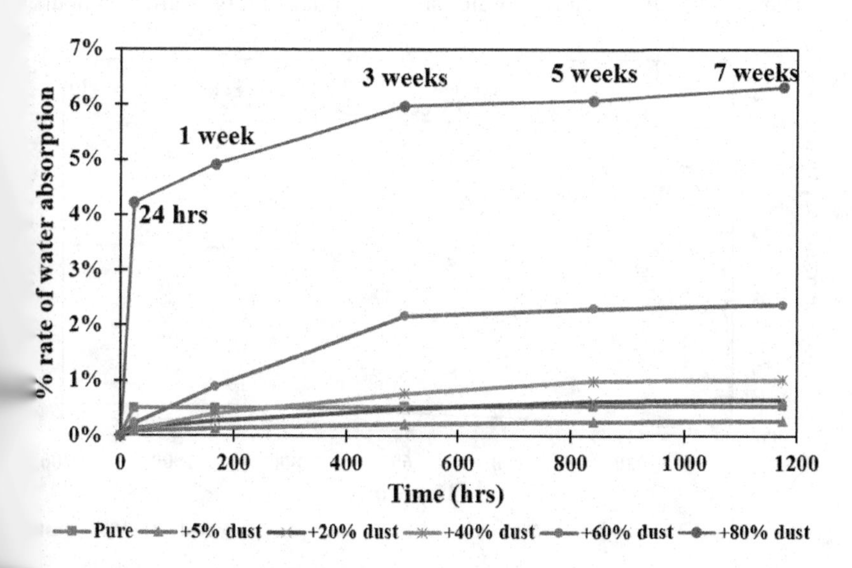

FIGURE 5.8 Water absorption rate at atmospheric pressure for PP–quarry dust composite prepared at different concentrations of quarry dust obtained from [10] under open access Creative Commons CC-BY.

reached after 24 hours at around 0.5%. In the case of using quarry dust as an additive at 5wt.% and 20wt.% composition, there is a positive effect on the water absorption rate. For the case of 5wt.% composition quarry dust, the water absorption rate is around 0.08% after 24 hours and saturation is reached after 3 weeks at 0.22wt.% This shows that quarry dust has a positive effect on the polymer matrix. In the case of 20wt.% quarry dust composition, the water absorption rate is around 0.13% after 24 hours and saturation is reached after 5 weeks at 0.6% which is still lower than that of pure PP. With a further increase in quarry dust composition (beyond 20wt.%), a high rate of water absorption is observed. In the case of 60wt.% quarry dust, the water absorption rate is at 0.2% after 24 hours and increases until saturation is reached after 5 weeks at 2.3%. For the case of 80% quarry dust, the water absorption rate is at 4.2% which increases at a faster rate until saturation is reached after 5 weeks at 6%. High rates of water absorption occur in samples containing 60wt.% and 80wt.% composition of quarry dust because of the large voids formed as the concentration of quarry dust is increased [44].

Figure 5.9 shows the water absorption rate for HDPE–quarry dust composites exposed at a pressure of 6 bars, and it is seen that the rate increases with an increase in the amount of quarry dust. In the case of 40wt.% quarry dust, the water absorption rate is at 0.1% after 24 hours and reaches saturation after 5 weeks at a water absorption rate of 0.75%. In the case of 60wt.% composition of quarry dust, there is a high water absorption rate at 0.65% after 24 hours

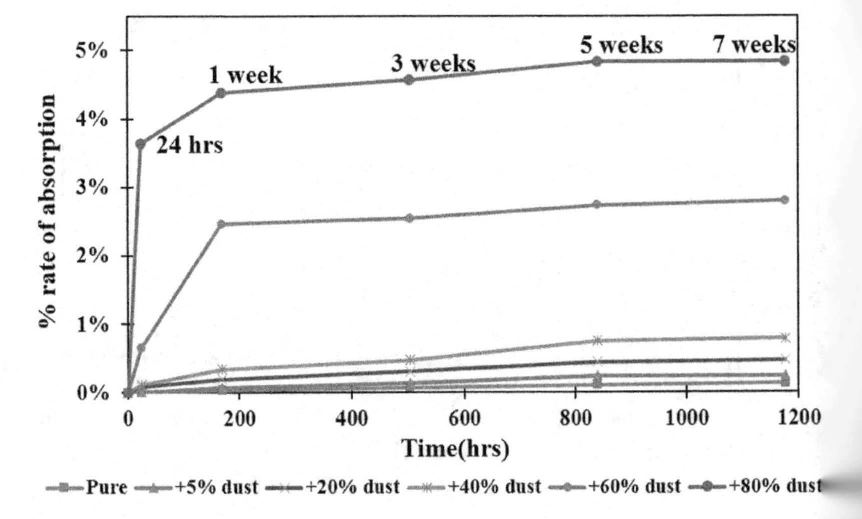

FIGURE 5.9 Water absorption rate for HDPE–quarry dust composite sample prepared at different concentrations of quarry dust exposed to a pressure of bars obtained from [10] under open access Creative Commons CC-BY.

and reaches saturation after 1 week at 2.5%. For 80wt.% quarry dust addition, after 24 hours, there is a very high water absorption rate of 3.6% which reaches saturation after 5 weeks at 4.8%. This is because the high amount of quarry dust in the composite weakens the bond between the quarry dust and the polymer matrix allowing more water to be absorbed in the resulting spaces.

However, comparing the results in Figures 5.7 and Figure 5.9, pure HDPE has the lowest water absorption rate when exposed to high-pressure condition (6 bars). This could be explained in terms of pressure-effect onto the composite structure; it is suggested that the high pressure compacts the pure HDPE leading to the closure of interfacial spaces (voids) such that there is minimal penetration of water molecules. By introducing quarry dust into the HDPE, the water absorption rate is reported higher at 6 bar pressure than at atmospheric conditions. The observation could similarly be attributed to the suggestion that high pressure weakens the interfacial bonding between the polymer matrix and the quarry dust by dislodging the quarry dust particles from their positions in the HDPE matrix, thereby forcing the water into the material.

Figure 5.10 shows the results of the water absorption rate test of PP–quarry dust samples carried out at a water pressure of 6 bars. As shown, the rate increases with an increase in the amount of quarry dust. In the case of pure PP, the water absorption rate increases with the increase in the amount of quarry dust at which the lowest water absorption is recorded at 0.04% after 24

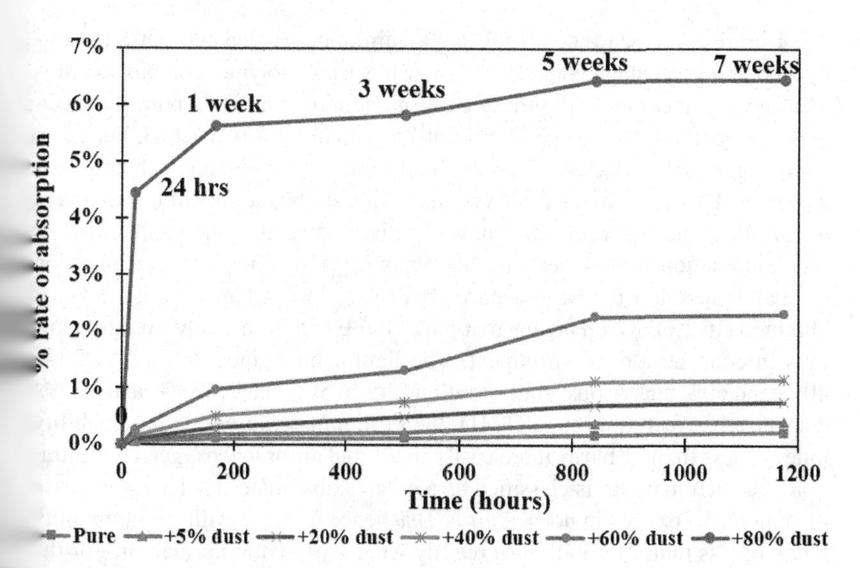

IGURE 5.10 Water absorption rate for PP–quarry dust composite samples preared at different concentrations of quarry dust exposed to a pressure of 6 bars btained from [10] under open access Creative Commons CC-BY.

hours and increases to a saturation point of 0.2% after 5 weeks. The low water absorption rate is because the pressure at 6 bars enhances compaction of pure PP structure as described earlier for HDPE samples. For the case of 40wt.% quarry dust, the water absorption rate increases from 0.15% to a saturation point of 1.12% after 5 weeks. This is due to the destruction of the bonding between the polymer matrix and the quarry dust, creating more spaces within the composite. For samples prepared at 60wt.% quarry dust composition, the water absorption rate is at 0.26% and increases up to a saturation point of 2.26% after 5 weeks whereas 80wt.% quarry dust samples have a water absorption rate of 4.4% and an increase in the saturation point of 6.4% after 5 weeks.

Similar to HDPE results, the water absorption rate of composite exposed to a pressure of 6 bars is higher when compared to samples exposed to atmospheric pressure. Just as explained in HDPE results, high pressure forces water into the composite increasing its water absorption capacity. Therefore, these results suggest that the composite cannot be used in high-pressure applications. Similar results have been reported for several composites exposed to hydraulic pressure to test their behaviour and long-term service under such conditions [45, 46].

5.5.4 Flammability Test (Limited Oxygen Index)

The Limiting Oxygen Index (LOI) is the minimum oxygen concentration calculated as a percentage volume necessary to sustain polymer combustion. It is measured by passing a mixture of oxygen and nitrogen over a burning specimen and reducing the oxygen level until a critical level is reached. For these composites to be effective in construction industry application, their stability and resistance to fire are important. In this study, flammability tests were undertaken, and the results are shown in Figure 5.11. It is shown that there is a linear relationship between the flammability index and the concentration of quarry dust. For the case of pure HDPE, the lowest flammability index is obtained (16.3%), which means that pure HDPE can burn easily even in a low oxygen concentration environment. The flammability indexes for 5%, 20% 40%, and 60% quarry dust concentrations are 16.5%, 17.2%, 17.4%, and 19.2% respectively. According to ASTMD 2863 [9], a material with a flammability index of less than 21 burns more easily in normal air or low oxygen concentra tion and such material is classified as a combustible material. This is becaus the amount of oxygen in normal air is 21% hence material with a flammabilit index of less than 21% will burn readily when exposed to normal air. For th case of 80% quarry dust, the flammability index is 24.4% which is more tha 21% and it means that the material cannot burn readily in normal air unless th oxygen concentration is increased to 24.4%. Composite materials containin

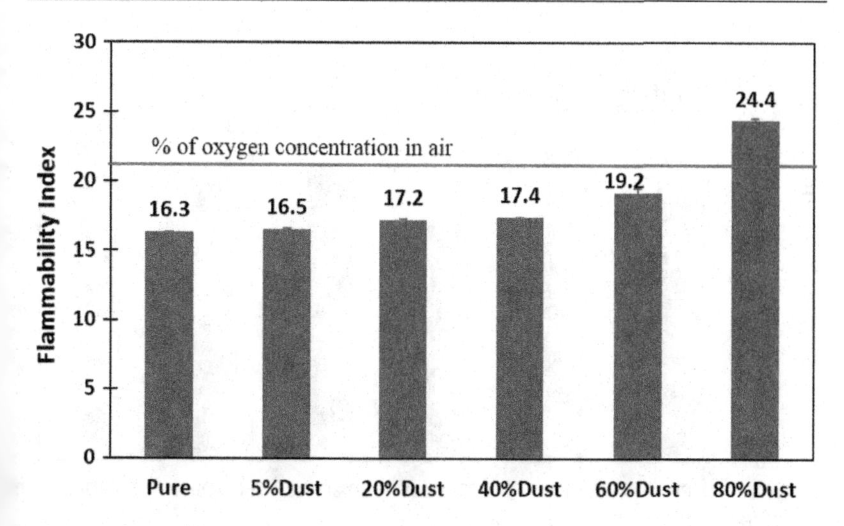

FIGURE 5.11 Limited Oxygen Index for HDPE–quarry dust composite samples prepared at different concentrations of the quarry dust obtained from [10] under open access Creative Commons CC-BY.

inorganic additive (such as quarry dust) tend to form char while burning [47]. The char acts as a flame retardant because its formation takes place at the expense of the volatile material, thereby increasing the level of oxygen necessary to sustain the flame. This is the reason why the flammability index increases as the amount of quarry dust increases. Therefore, HDPE containing 80% of quarry dust is best suited for insulation purposes and can be industrially applied in roofing in an environment having relatively high temperatures and risks of fires.

Figure 5.12 shows the flammability of the composite material prepared from the PP matrix. The flammability index increases as the amount of quarry dust increase due to an increase in the formation of char from the combustion of inorganic components which inhibit the flammability of the polymer. This means that quarry dust can act as a flame-retardant material in PP in applications that need properties of PP and at the same time in an environment that is exposed to a relatively higher temperature than that at which pure PP can ignite. Generally, it is observed that the composites made from PP and quarry dust have lower flammability indices than 21, which means that all the composites containing PP will burn readily in normal air. Hence, the application of PP–quarry dust composites is not recommended in fire-risk environments such as in forests and industrial buildings, to mention but a few. However, flame-retardant additives can be used in the PP–quarry dust composite to boost its flammability index. Some of the additives include layered double hydroxides

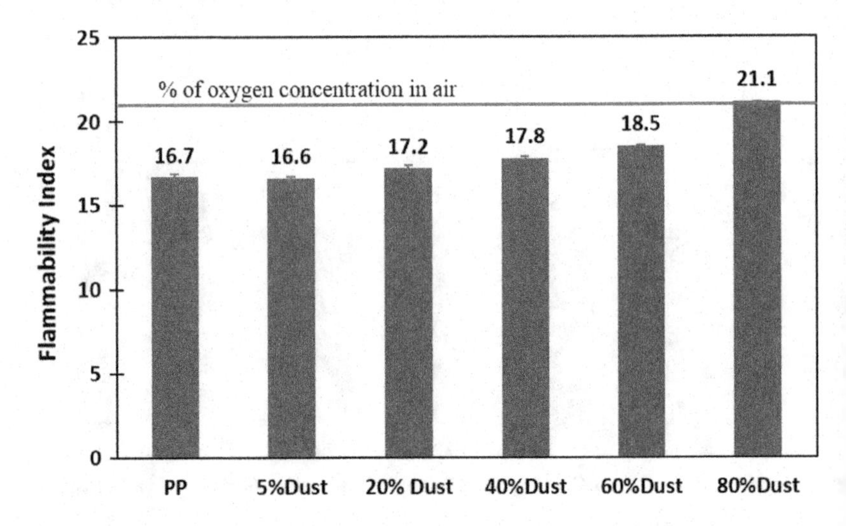

FIGURE 5.12 Limited Oxygen Index for PP–quarry dust composite samples prepared at different concentrations of the quarry dust obtained from [10] under open access Creative Commons CC-BY.

(LDHs) [48], ethylene-vinyl acetate [49], and CeO_2-dolomite [50], to mention but a few.

5.5.5 Thermal Analysis

Measurements obtained from a differential scanning calorimetry are used to study the crystallization and melting behaviour and the oxidative and thermal stability of a sample. The crystalline melt temperature, glass transition temperature, and the specific heat capacity of a sample can be determined [51].

Thermal analysis was conducted on the prepared composite at different quarry dust composition using differential scanning calorimetry to investigate the crystallization and melting behaviour of the composites. The results obtained from this test were the crystallinity index (X_c), melting temperature (T_m), and the glass transition temperature (T_g). The temperatures of the peaks were taken to be the melting temperatures of the DSC melting endotherm [52]. The nanophase changes in a material are well demonstrated by the glass transition temperature.

Figure 5.13 shows the DSC thermogram results for pure HDPE, 5%, 20%, 40%, 60%, and 80% quarry dust composites. One endothermic peak was observed in each of the curves, and this is attributed to the primary nucleation of crystals or the formation of a new crystalline phase [53]. The results indicat

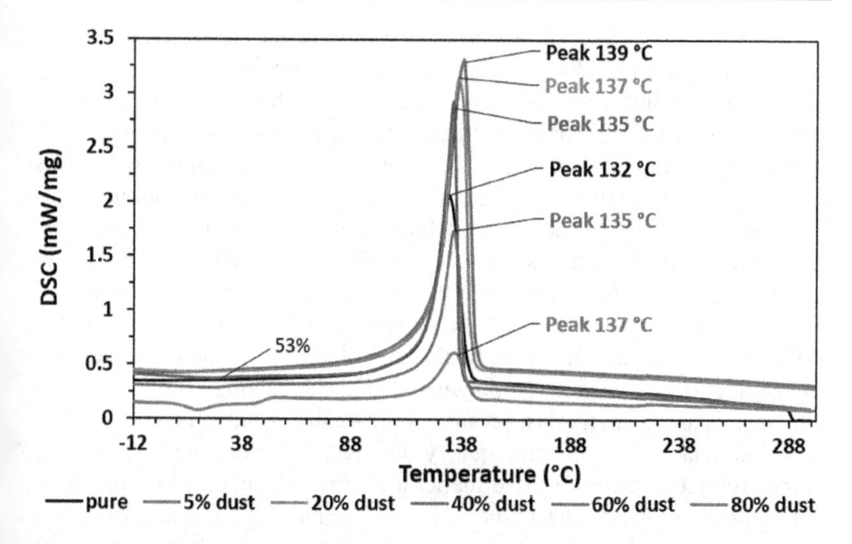

FIGURE 5.13 Differential scanning calorimetry (DSC) thermograms of HDPE–quarry dust composite samples prepared at different concentrations of the quarry dust

that the material has a degree of crystallinity of 53% in the case of pure HDPE material. With the use of quarry dust as an additive to the HDPE, the crystallization point is not captured by the DSC but it is suspected to have increased. This is because quarry dust is a ceramic material that contains majorly silica, and it is expected to be crystalline, making the composite's degree of crystallinity increase. All the samples have a melting peak of between 132°C and 139°C. The high melting temperature peaks are attributed to the HDPE matrix having few chain branches making the chain crystallize at high temperature or the chains having a high molecular weight [13]. The HDPE composite with 5wt.% quarry dust composition has the highest melting peak temperature of 139°C. The melting peak temperature decreases as the composition of quarry dust is increased up to 40wt.% composition and starts to increase with a further increase in the composition of quarry dust. The presence of agglomeration may have affected the sampling for the DSC test where the Tm obtained were anomalous. Pure HDPE has the least melting peak temperature, and this is could be due to high chain branched in the molecular structure unlike where quarry dust has been added as a reinforcement agent, reducing the number of chain branches. In the case of 80wt.% concentration of quarry dust in HDPE, two shoulders are observed at 20°C and 50°C which are attributed by the difference crystallization cooling rates. The energy in the melting peak decreases with an increase in the amount of quarry dust due to the small amount of the polymer matrix. The shapes of the curves are observed to be the same and this means that adding quarry dust does not change the thermal properties of the material.

Figure 5.14 shows thermograms obtained through a DSC test of plastic–quarry dust composite prepared in different weight proportions of quarry dust. It is observed that the curves have two different glass transition temperatures. This is attributed to the ability of polypropylene to form different crystalline points in the polymer structure due to the cooling phenomenon of PP and it can build different crystalline structures [13]. Co-crystallization can be caused by irregularities within the crystal structure [13]. The crystalline point is also influenced by the addition of quarry dust. All the samples have a melting peak of between 163°C and 170°C, each influenced by the level of agglomeration in the prepared sample due to variations in the polymer chain. The melting energy decreases with an increase in quarry dust because of the small amount of polymer matrix.

Additionally, PP – 80wt.% quarry dust composite has the least melting temperature, T_m, making it have the least energy absorbed as the composite material is being heated due to the high composition of quarry dust. The presence of quarry dust tends to reduce the number of polymer chains that need to be broken down; hence less thermal energy is needed to cause melting, unlike where the polymer chains are long. This means that for pure PP, energy must be increased so that it can achieve its melting peak. Generally, the melting temperatures of composites made of HDPE are lower than those made up of PP. This means that composites made of HDPE have fewer polymer chains and molecular weight as compared to composites made up of PP.

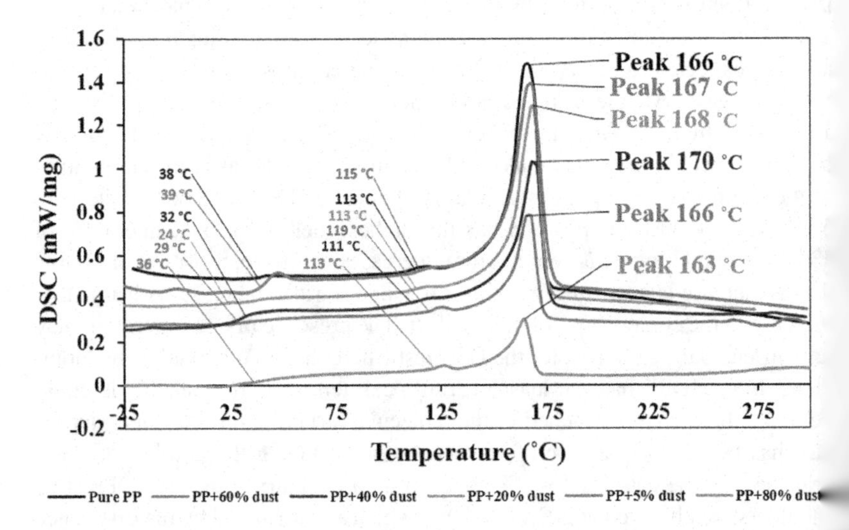

FIGURE 5.14 DSC test for conduction thermal analysis on the quarry dust–P composite samples prepared at different concentrations of the quarry dust.

5.5.6 Fourier Transformed Infrared Characterization

This characterization technique is used to confirm the material structure of the composite made up of polymer and quarry dust (silica). The assignment of the wavenumbers obtained from the analysis is summarized in Table 5.3. Figure 5.15 shows the FTIR spectrum of the pure PP sample obtained.

This assignment confirms the presence of pure polymeric material. The peaks at 2917 cm^{-1}, 2951 cm^{-1}, 2838 cm^{-1}, 1376 cm^{-1}, and 1456 cm^{-1} are very strong, and this is attributed to the stretching and bending of the polypropylene molecular chains.

Silicates are usually known to have wavenumber in the range 400–1250 cm^{-1}. Additionally, asymmetric Si-O-Si stretching vibrations occur in the range 830–1250 cm^{-1}, with the O-Si-O deformation or bending modes occurring in the range 400 cm^{-1}–560 cm^{-1} and symmetric Si-O-Si stretching vibrations occurring in the weaker peaks in the range 670 cm^{-1}–830 cm^{-1} [54]. Figure 5.16 shows the FTIR spectrum for PP–quarry dust at 80wt.% in concentration. The peaks observed at 997 cm^{-1} are attributed to asymmetric Si-O-Si stretching vibrations. The weak peaks observed at 627 cm^{-1}, 678 cm^{-1}, and 748 cm^{-1} are attributed to symmetric Si-O-Si stretching vibrations while the band observed at 426 cm^{-1}, 449 cm^{-1}, and 509 cm^{-1} are attributed to O-Si-O deformation or bending modes. The band at 3698 cm^{-1} is attributed to the formation of an O-H group.

This band confirms the hydrophilic nature of the composite. The significant spectra peak is 912 cm^{-1}, which depicts an illite group. Illite is a micaceous mineral that is non-expanding, usually found in sedimentary rock and

TABLE 5.3 Various Assignments of Different Wavenumbers of the Resulting Bands in PP

WAVENUMBER (CM^{-1})	*ASSIGNMENT*
2951	C-H stretch
2917	C-H stretch
2838	C-H stretch
1456	CH_2 bend
1376	CH_3 bend
1166	CH_3 rock, CH bend, and C-C stretch
997	CH_3 rock, CH bend, and CH_3 bend
973	CH_3 rock and C-C stretch
841	CH_2 rock and C-CH_3 stretch
808	CH_2 rock and C-C stretch

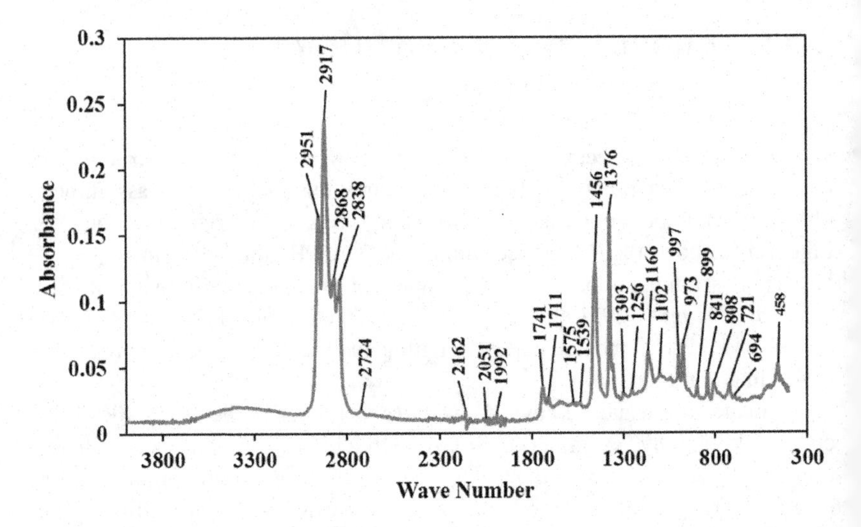

FIGURE 5.15 Fourier Transformed Infrared (FTIR) spectroscopy of pure PP sample showing various peaks in the spectrum.

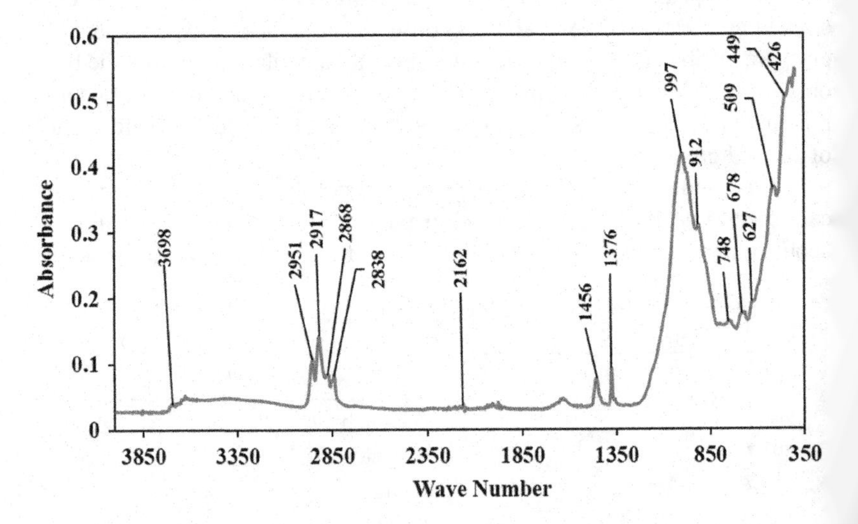

FIGURE 5.16 Fourier Transformed Infrared (FTIR) spectroscopy of quarry dust-PP composite sample prepared at 80wt.% concentrations of the quarry dust.

soil, formed in weathering and hydrothermal environments [55]. Some band observed in the pure PP spectrum are also observed in the spectrum of PP - 80wt.% quarry dust composite. These bands are 2951 cm^{-1}, 2917 cm^{-1}, 286 cm^{-1}, and 2838 cm^{-1} attributed to C-H stretching and 1456 cm^{-1} and 1376 cm which are attributed to C-H bending. However, these bands are weak when compared to the pure PP spectrum. This means that the polymer concentration has reduced, hence the magnitude of stretching and bending also reduces.

The FTIR shown in Figure 5.17 shows a peak around 2914 cm^{-1} and a second peak around 2844 cm^{-1} which indicate that the sample has methyl as one functional group and two main peaks of around 1471 cm^{-1} and 717 cm^{-1} indicate that sample has methylene as another functional group. Given the functional group, it can be concluded that the structure of this polymer must be polyethene. The FTIR shows a typical polyethene spectrum. The assignment is summarized in Table 5.4.

From Figure 5.18, various peaks are observed with the same stretching in the range 400–1200 cm^{-1} observed in both PP + 80%wt. and HDPE + 80wt.%. Each wavenumber is assigned to a functional group. As previously explained, the peaks observed at 997 cm^{-1} are attributed to asymmetric Si-O-Si stretching vibrations. The weak peaks observed at 627 cm^{-1}, 678 cm^{-1}, and 748 cm^{-1} are due to symmetric Si-O-Si stretching vibrations and the peaks at 426 cm^{-1}, 449 cm^{-1}, and 509 cm^{-1} are attributed to O-Si-O deformation or bending modes. The peaks at the broad absorption band of 3440–3700 cm^{-1} and 1633 cm^{-1} are due to the –OH groups. This show the composite is hydrophilic in nature [56]. Bands are also observed at 2914 cm^{-1} and 2846 cm^{-1} which are attributed to C-H stretching and at 1471 cm^{-1}, 1377 cm^{-1}, and 717 cm^{-1} which are attributed to CH_2 bend and rock [57]. However, the intensity of these peaks is less than those observed in the pure HDPE spectrum indicating that the composition of

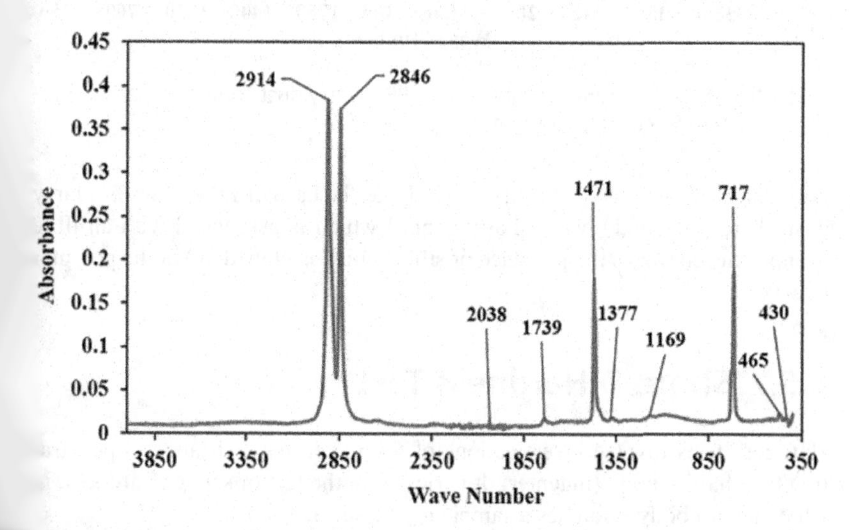

FIGURE 5.17 Fourier Transformed Infrared (FTIR) spectroscopy of pure HDPE sample showing various peaks in the spectrum.

TABLE 5.4 Various Assignments of Different Wavenumbers of the Resulting Bands in HDPE

WAVENUMBER (CM^{-1})	*ASSIGNMENT*
2914	C-H stretch
2844	C-H stretch
1471	CH_2 bend
1377	CH_2 bend
717	CH_2 rock

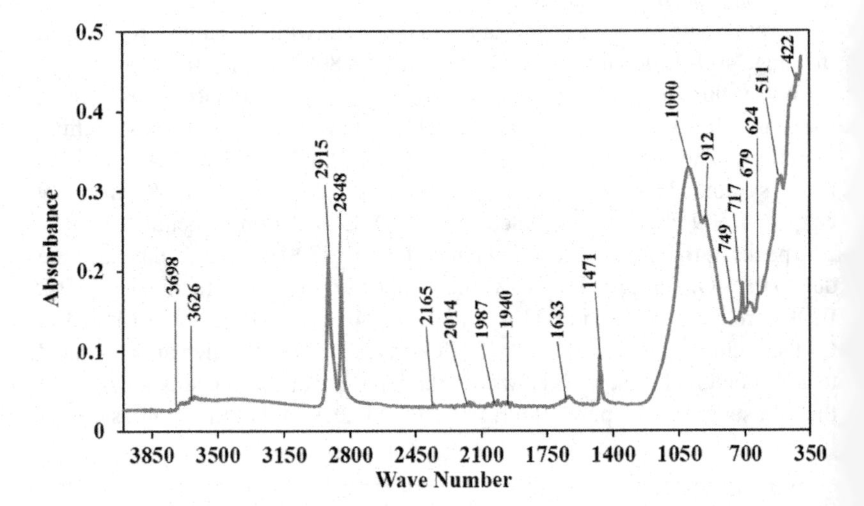

FIGURE 5.18 FTIR characterization of HDPE–quarry dust composite sample prepared at 80wt.% concentration of the quarry dust.

polymeric material has reduced. Similarly, to the case of PP – 80wt.% quarry dust, there is a band observed at 912 cm^{-1} which is associated with an illite group. This confirms the presence of silica which is embedded in the polymer matrix

5.5.7 Shore D Hardness Test

Hardness tests involve investigations of the material's resistance to penetration by a harder body (indenter). It varies from the test of scratch hardness, in which a hard body scratches a sample on the surface.

Shore D Hardness test is important for materials used in construction applications because it indicates the resistance of such material to scratch from

TABLE 5.5 Effects of Quarry Dust Content on the Hardness of Plastic–Quarry Dust Composite Samples Prepared at Different Concentrations of Quarry Dust

	SHORE D HARDNESS						
MATERIAL	*TRIAL 1*	*TRIAL 2*	*TRIAL 3*	*TRIAL 4*	*TRIAL 5*	*AVERAGE*	*STD. ERROR*
PP	72	71	68	71	76	71.6	1.2884
PP + 5% dust	72	77	74	75	72	74	0.9487
PP + 20% dust	76	72	76	75	76	75	0.7746
PP + 40% dust	78	78	80	80	77	78.6	0.6000
PP + 60% dust	74	83	82	80	79	79.6	1.5684
PP + 80% dust	83	81	85	81	82	82.4	0.7483
HDPE	63	66	67	67	70	66.6	1.1224
HDPE + 5% dust	71	66	63	70	68	67.6	1.4353
HDPE + 20% dust	70	73	72	71	71	71.4	0.5099
HDPE + 40% dust	75	73	72	73	74	73.4	0.5099
HDPE + 60% dust	76	72	75	77	77	75.4	0.9274
HDPE + 80% dust	79	83	84	81	83	82	0.8944

external sliding loads. Table 5.5 shows the results of Shore D Hardness tests for the plastic composites fabricated in this study. As illustrated, using quarry dust as an additive in the polymer matrix shows a positive effect in both HDPE and PP polymers. In the case of HDPE, the Shore D Hardness increases linearly from 66 for pure HDPE to 82 for 80wt.% quarry dust concentration. The same is observed in PP where the Shore D Hardness increases from 71 for pure PP to 82 for 80wt.% quarry dust concentration. These results are obvious and as expected since the quarry dust was obtained from hard rocks and as such are likely to increase the hardness of the composite as its percentage composition in the composite is increased. However, the hardness of PP is much higher than that of HDPE due to the high hardness of PP as compared to HDPE. In the case of using 80wt.% quarry dust composition, the composite of PP and HDPE matrix have the same Shore D Hardness value of 82. This is an interesting finding, implying that the easier to process HDPE can be improved to match the hardness of harder-to-process PP polymer-based composite.

5.6 SUMMARY

In this chapter, preparation and characterization of a typical polymer-silica composite is presented as a case study. The samples were prepared by mixing pure PP and HDPE with varying amounts of quarry dust. The test samples were then obtained via compression moulding. The samples were prepared with quarry dust amounts of 0%, 5%, 20%, 40%, 60%, and 80% to evaluate its influence on the composite performance. The composites were characterized for microscopy, chemical properties, thermal behaviour, water absorption, flammability characteristics, and hardness. It was discovered that it is indeed possible to prepare plastic–quarry dust composite for construction applications. It was also reported that various characteristics of the polymer composite were influenced by the quantity of quarry dust. Based on the results presented, the HDPE/PP–quarry dust can be used in various areas of construction such as roofing tiles, pavement blocks, interior wall tiling, and wall cladding. Further evaluations can be undertaken to explore more applications of the composite.

REFERENCES

1. S. Ettlinger, E. Tim, B. Katharine, K. Stephen, C. Wangari, and N. Eve, "Plastic packaging waste flows in Kenya," *Danish Environmental Protection Agency*, 2018. [Online]. Available: www.eunomia.co.uk
2. K. B. Adhikary, S. Pang, and M. P. Staiger, "Dimensional stability and mechanical behaviour of wood–plastic composites based on recycled and virgin high-density polyethylene (HDPE)," *Composites Part B: Engineering*, vol. 39, no. 5, pp. 807–815, 2008, doi: 10.1016/j.compositesb.2007.10.005.
3. L. Gustavo Barbosa, M. Piaia, and G. Henrique Ceni, "Analysis of impact and tensile properties of recycled polypropylene," IJME, vol. 7, no. 6, pp. 117–120, 2017, doi: 10.5923/j.ijme.20170706.03.
4. T. Ahmed and O. Mamat, "The development and characterization of HDPE-silica sand nanoparticles composites," in *2011 IEEE Colloquium on Humanities, Science and Engineering*, Penang, Malaysia, 2011, pp. 6–11.
5. D. A. Hansen and R. B. Puyear, *Materials Selection for Hydrocarbon and Chemical Plants*. New York: Marcel Dekker, 1996.
6. www.iso.org, ISO 179-1:2010(en), *Plastics – Determination of Charpy Impact Properties – Part 1: Non-Instrumented Impact Test*. [Online]. Available: https:// www.iso.org/obp/ui/fr/ (accessed February 18, 2020).
7. R. J. Crawford, Ed., *Plastics Engineering*, 3rd ed. Oxford: Butterworth Heinemann, 1998.
8. ASTM D-570-98, *Test Method for Water Absorption of Plastics*. West Conshohocken, PA: ASTM International, 1999, www.astm.org
9. ASTM D2863-19, *Standard Test Method for Measuring the Minimum Oxygen Concentration to Support Candle-Like Combustion of Plastics (Oxygen Index)*. West Conshohocken, PA: ASTM International, 2019, www.astm.org
10. H. Shagwira, F. Mwema, T. Mbuya, and A. Adediran, "Dataset on impact strength, flammability test and water absorption test for innovative

polymer-quarry dust composite," *Data in Brief*, vol. 29, p. 105384, 2020, doi: 10.1016/j.dib.2020.105384.

11. www.iso.org/obp, ISO 11357-1:2016(en), *Plastics – Differential Scanning Calorimetry (DSC) – Part 1: General Principles*. [Online]. Available: https://www.iso.org/obp/ui/ (accessed February 18, 2020).
12. ASTM D2240-15(2021), Standard Test Method for Rubber Property—Durometer Hardness, ASTM International, West Conshohocken, PA, 2021, www.astm.org
13. A. Frick, C. Stern, and V. Muralidharan, *Practical Testing and Evaluation of Plastics*. Weinheim, Germany: Wiley-VCH Verlag GmbH & Co, 2019.
14. R. W. Rice, *Mechanical Properties of Ceramics and Composites: Grain and Particle Effects/Roy W. Rice*. New York: Marcel Dekker, 2000.
15. G. Wu, Q. Zhang, X. Yang, Z. Huang, and W. Sha, "Effects of particle/matrix interface and strengthening mechanisms on the mechanical properties of metal matrix composites," *Composite Interfaces*, vol. 21, no. 5, pp. 415–429, 2014, doi: 10.1080/15685543.2014.872914.
16. Y. Xiang, D. J. Srolovitz, L.-T. Cheng, and E. Weinan, "Level set simulations of dislocation-particle bypass mechanisms," *Acta Materialia*, vol. 52, no. 7, pp. 1745–1760, 2004, doi: 10.1016/j.actamat.2003.12.016.
17. G. Feng, X. Wang, Y. Kang, and Z. Zhang, "Effect of thermal cycling-dependent cracks on physical and mechanical properties of granite for enhanced geothermal system," *International Journal of Rock Mechanics and Mining Sciences*, vol. 134, no. 5, p. 104476, 2020, doi: 10.1016/j.ijrmms.2020.104476.
18. R. Strapasson, S. C. Amico, M. F. R. Pereira, and T. H. D. Sydenstricker, "Tensile and impact behavior of polypropylene/low density polyethylene blends," *Polymer Testing*, vol. 24, no. 4, pp. 468–473, 2005, doi: 10.1016/j.polymertesting.2005.01.001.
19. W. Gerberich and W. Yang, "Interfacial and nanoscale failure," in *Comprehensive Structural Integrity, I*, Milne, R. O. Ritchie and B. L. Karihaloo, Eds. Amsterdam and Boston, MA: Elsevier/Pergamon, 2003, pp. 1–40.
20. J. E. Perez Ipiña and C. Berejnoi, "Analysis of specimen size conversion in the ductile to brittle transition region of ferritic steels," *Procedia Structural Integrity*, vol. 2, pp. 769–776, 2016, doi: 10.1016/j.prostr.2016.06.099.
21. S.-Y. Fu, X.-Q. Feng, B. Lauke, and Y.-W. Mai, "Effects of particle size, particle/matrix interface adhesion and particle loading on mechanical properties of particulate–polymer composites," *Composites Part B: Engineering*, vol. 39, no. 6, pp. 933–961, 2008, doi: 10.1016/j.compositesb.2008.01.002.
22. W. D. Callister, *Materials Science and Engineering: An Introduction/William D. Callister, Jr*, 7th ed. New York and Chichester: Wiley, 2007.
23. M. Guagliano and M. H. Aliabadi, *Fracture and Damage of Composites*. Ashurst: WIT Press, 2005. [Online]. Available: https://ebookcentral.proquest.com/lib/gbv/detail.action?docID=511991
24. A. Paknia, A. Pramanik, A. R. Dixit, and S. Chattopadhyaya, "Effect of size, content and shape of reinforcements on the behavior of metal matrix composites (MMCs) under tension," *Journal of Materials Engineering and Performance*, vol. 25, no. 10, pp. 4444–4459, 2016, doi: 10.1007/s11665-016-2307-x.
25. A. W. Batchelor, L. N. Lam, and M. Chandrasekaran, *Materials Degradation and Its Control by Surface Engineering*, 2nd ed. London: Imperial College Press, 2002.

26. J. Mathew, G. Remy, M. A. Williams, F. Tang, and P. Srirangam, "Effect of Fe intermetallics on microstructure and properties of Al-7Si alloys," *JOM*, vol. 71, no. 12, pp. 4362–4369, 2019, doi: 10.1007/s11837-019-03444-5.
27. O. Gebremeskel, A. Thakur, and M. Mariam, "Crack propagation analysis of fiber-reinforced composite hollow transmission shaft," *International Journal of Current Engineering and Technology*, Vol. 8, No.3, pp. 698–707, 2018, http://inpressco.com/category/ijcet
28. D. Peretz and A. T. DiBenedetto, "Crack propagation in polymeric composites," *Engineering Fracture Mechanics*, vol. 4, no. 4, pp. 979–990, 1972, doi: 10.1016/0013-7944(72)90029-X.
29. T. Kawaguchi and R. A. Pearson, "The moisture effect on the fatigue crack growth of glass particle and fiber reinforced epoxies with strong and weak bonding conditions: part 1. Macroscopic fatigue crack propagation behavior," *Composites Science and Technology*, vol. 64, no. 13, pp. 1981–1989, 2004, doi: 10.1016/j.compscitech.2004.02.016.
30. M. Jawaid, M. Thariq, and N. Saba, *Failure Analysis in Biocomposites, Fibre-Reinforced Composites and Hybrid Composites*. Oxford: Woodhead Publishing, 2018.
31. P. S. Theocaris and C. A. Stassinakis, "Crack propagation in fibrous composite materials studied by SEM," *Journal of Composite Materials*, vol. 15, no. 2, pp. 133–141, 1981, doi: 10.1177/002199838101500203.
32. V.N. Hristov, R. Lach, and W. Grellmann, "Impact fracture behavior of modified polypropylene/wood fiber composites," *Polymer Testing*, vol. 23, no. 5, pp. 581–589, 2004, doi: 10.1016/j.polymertesting.2003.10.011.
33. C.-W. Hsu, L. Wang, and W.-F. Su, "Effect of chemical structure of interface modifier of TiO2 on photovoltaic properties of poly(3-hexylthiophene)/TiO2 layered solar cells," *Journal of Colloid and Interface Science*, vol. 329, no. 1, pp. 182–187, 2009, doi: 10.1016/j.jcis.2008.10.008.
34. K. Katueangngan, T. Tulyapitak, A. Saetung, S. Soontaranon, and N. Nithi-uthai, "Renewable interfacial modifier for silica filled natural rubber compound," *Procedia Chemistry*, vol. 19, pp. 447–454, 2016, doi: 10.1016/j.proche.2016.03.037.
35. V. M. Gun'ko, I. N. Savina, and S. V. Mikhalovsky, "Properties of water bound in hydrogels," *Gels (Basel, Switzerland)*, vol. 3, no. 4, pp. 1–30, 2017, doi: 10.3390/gels3040037.
36. N. S. Othman, R. Santiagoo, Z. Mustafa, W. A. Mustafa, I. Zunaidi, W. K. Wan, Z. M. Razlan, and A. B. Shahriman, "Studies on water absorption of polypropylene/recycled acrylonitrile butadiene rubber/empty fruit bunch composites," *IOP Conference Series: Materials Science and Engineering*, vol. 429, p. 12091, 2018 doi: 10.1088/1757-899X/429/1/012091.
37. M. Karimi, "Diffusion in polymer solids and solutions," in *Mass Transfer in Chemical Engineering Processes*, J. Marko, Ed., pp. 17–40. London: InTechOpen Limited, 2011.
38. E. T. Mbou, E. Njeugna, A. Kemajou, N. R. T. Sikame, and D. Ndapeu "Modelling of the water absorption kinetics and determination of the water diffusion coefficient in the pith of *Raffia vinifera* of Bandjoun, Cameroon, *Advances in Materials Science and Engineering*, vol. 2017, no. 13, pp. 1–12 2017, doi: 10.1155/2017/1953087.

39. A. Akashi, Y. Matsuya, M. Unemori, and A. Akamine, "The relationship between water absorption characteristics and the mechanical strength of resin-modified glass-ionomer cements in long-term water storage," *Biomaterials*, vol. 20, no. 17, pp. 1573–1578, 1999, doi: 10.1016/S0142-9612(99)00057-5.
40. H. Deng, C. T. Reynolds, N. O. Cabrera, N.-M. Barkoula, B. Alcock, and T. Peijs, "The water absorption behaviour of all-polypropylene composites and its effect on mechanical properties," *Composites Part B: Engineering*, vol. 41, no. 4, pp. 268–275, 2010, doi: 10.1016/j.compositesb.2010.02.007.
41. M. Jawaid, M. Thariq, and N. Saba, *Mechanical and Physical Testing of Biocomposites, Fibre-Reinforced Composites and Hybrid Composites*. Oxford: Woodhead Publishing, 2018.
42. A. Naceri, "An analysis of moisture diffusion according to Fick's law and the tensile mechanical behavior of a glass-fabric-reinforced composite," *Mechanics of Composite Materials*, vol. 45, no. 3, pp. 331–336, 2009, doi: 10.1007/s11029-009-9080-y.
43. A. M. Radzi, S. M. Sapuan, M. Jawaid, and M. R. Mansor, "Water absorption, thickness swelling and thermal properties of roselle/sugar palm fibre reinforced thermoplastic polyurethane hybrid composites," *Journal of Materials Research and Technology*, vol. 8, no. 5, pp. 3988–3994, 2019, doi: 10.1016/j.jmrt.2019.07.007.
44. M. R. Sanjay and B. Yogesha, "Study on water absorption behaviour of jute and kenaf fabric reinforced epoxy composites: hybridization effect of E-glass fabric," *International Journal of Composite Materials*, vol. 6, no. 2, pp. 55–62, 2016.
45. S. Chen, H. Xu, H. Duan, M. Hua, L. Wei, H. Shang, and J. Li, "Influence of hydrostatic pressure on water absorption of polyoxymethylene: experiment and molecular dynamics simulation," *Polymers for Advanced Technologies*, vol. 28, no. 1, pp. 59–65, 2017, doi: 10.1002/pat.3858.
46. C. Li, G. Xian, and H. Li, "Water absorption and distribution in a pultruded unidirectional carbon/glass hybrid rod under hydraulic pressure and elevated temperatures," *Polymers*, vol. 10, no. 6, p. 627, 2018, doi: 10.3390/polym10060627.
47. A. Pattiya, "1 – Fast pyrolysis," in *Direct Thermochemical Liquefaction for Energy Applications*, L. Rosendahl, Ed. Oxford: Woodhead Publishing, 2017, pp. 3–28. [Online]. Available: https://www.sciencedirect.com/science/article/pii/B9780081010297000011
48. K. Shanmuganathan and C. J. Ellison, "Layered double hydroxides," in *Polymer Green Flame Retardants*, C. D. Papaspyrides and P. Kiliaris, Eds., Amsterdam, Netherlands: Elsevier, 2014, pp. 675–707.
49. L. Geoffroy, F. Samyn, M. Jimenez, and S. Bourbigot, "Additive manufacturing of fire-retardant ethylene-vinyl acetate," *Polymers for Advanced Technologies*, vol. 30, no. 7, pp. 1878–1890, 2019, doi: 10.1002/pat.4620.
50. J. Zoleta, G. Itao, V. J. Resabal, A. Lubguban, R. Corpuz, C. Tabelin, M. Ito, and N. Hiroyoshi, "CeO 2 -dolomite as fire retardant additives on the conventional intumescent coating in steel substrate for improved performance," *MATEC Web of Conferences*, vol. 268, no. 2, p. 4009, 2019, doi: 10.1051/matecconf/201926804009.
51. P. Gill, T. T. Moghadam, and B. Ranjbar, "Differential scanning calorimetry techniques: applications in biology and nanoscience," *Journal of Biomolecular Techniques: JBT*, vol. 21, no. 4, pp. 167–193, 2010.

52. M. M. Ibrahim, A. Dufresne, W. K. El-Zawawy, and F. A. Agblevor, "Banana fibers and microfibrils as lignocellulosic reinforcements in polymer composites," *Carbohydrate Polymers*, vol. 81, no. 4, pp. 811–819, 2010, doi: 10.1016/j.carbpol.2010.03.057.
53. A. J. Varma, S. V. Deshpande, and J. F. Kennedy, "Metal complexation by chitosan and its derivatives: a review," *Carbohydrate Polymers*, vol. 55, no. 1, pp. 77–93, 2004.
54. C. M. Müller, A. Molinelli, M. Karlowatz, A. Aleksandrov, T. Orlando, and B. Mizaikoff, "Infrared attenuated total reflection spectroscopy of quartz and silica micro- and nanoparticulate films," *The Journal of Physical Chemistry C*, vol. 116, no. 1, pp. 37–43, 2012, doi: 10.1021/jp205137b.
55. A. Chandrasekaran, A. Rajalakshmi, R. Ravisankar, and S. Kalarasai, "Analysis of beach rock samples of Andaman Island, India by spectroscopic techniques," *Egyptian Journal of Basic and Applied Sciences*, vol. 2, no. 1, pp. 55–64, 2015, doi: 10.1016/j.ejbas.2014.12.004.
56. A. Thongphud, P. Visal-athaphand, P. Supaphol, and B. Paosawatyanyong, "Improvement of hydrophilic properties on electrospun polyacrylonitrile fabrics surface by plasma treatment," *Advanced Materials Research*, vol. 213, pp. 103–106, 2011, doi: 10.4028/www.scientific.net/AMR.213.103.
57. W. F. Maddams, "A review of Fourier-transform Raman spectroscopic studies on polymers," *Spectrochimica Acta Part A: Molecular Spectroscopy*, vol. 50, no. 11, pp. 1967–1986, 1994, doi: 10.1016/0584-8539(94)80209-2.

Future Outlook 6

6.1 INTRODUCTION

It is no doubt that the demand and market for composite materials will continue to increase. The future of the polymeric composite market is attractive due to expanding opportunities in various sectors such as consumer goods, energy, transportation, electronics, and others. For instance, it is estimated that the composite market would grow by over $40 billion by 2024 [1]. In all sectors of modern society, there is an increasing demand for miniature and lightweight materials. Additionally, there is high demand for smart gadgets, portable energy sources, powerful machines, flexible gadgets, and so forth. These factors have created a necessity of coming up with composite materials that will tend to satisfy these demands. The need for better and advanced composite materials will significantly be enhanced by Industry 4.0. The technologies driving Industry 4.0 include: Internet of Things (IoT), automation, simulation, 3D printing, and analytics. Most sectors like the industries and the government are continuously adopting these technologies to boost their efficiencies, improve the quality of their services, and develop new operating models which can respond to market unpredictability.

In this chapter, an outlook of the application of composite materials in construction and related industries is briefly discussed. In specific, the chapter forecasts the future of application and need for composite materials in Industry 4.0 and the emerging materials.

6.2 COMPOSITES AND INDUSTRY 4.0 IN CONSTRUCTION INDUSTRY

The construction industry is rapidly revolutionizing due to (i) demand for low-cost housing, (ii) the need for safety against natural calamities brought

DOI: 10.1201/9781003231936-6

about by climatic changes, (iii) the need for comfort, and (iv) and enhanced security. In the future, due to climate change, there is a likelihood of the increase of collapse of structures. Climate change brings about water encroachment and flooding, storms, snow, and increased temperatures [2]. Therefore, the modern and future construction industry should adopt technologies that can enhance the adaption of buildings and structures to climate change. One of the focuses of these technologies should be on the development of new and sustainable materials for the construction industry. Polymeric composites are candidate materials for the sustainable construction industry. In general, modern buildings and structures should have the following features:

i. They should be structurally safe such that they can withstand fires, earthquakes, storms, etc.
ii. They should be comfortable and should not pose health risks to humans. The buildings should be able to maintain comfortable and clean conditions for human habitat.
iii. The modern construction industry demands the rapid construction of structures and buildings. For instance, in cases of natural calamities, governments and other bodies need instant technologies to put up structures for the affected persons.
iv. The buildings for the future demand the incorporation of smart systems for security and control due to the complex nature of the application of technology in crime.

To satisfy these features, the type and properties of construction will play a key role. The technologies adopted in construction are strongly dependent on the construction materials employed. One of the most exciting technologies, which can revolutionize the construction sector is additive manufacturing (commonly referred to as 3D printing). 3D printing technology involves the creation of three-dimensional components/structures (with the use of computer control) by combining raw materials through a layer-by-layer procedure [3]. In considering the construction sector, this technology can be evaluated in a twofold perspective: (i) development of materials for 3D printing in the construction industry, and (ii) possibility of direct 3D printing of structures and buildings. The technology would offer the following benefits to the modern construction sector [4]:

i. Onsite material production and assembly – with 3D printing, the construction materials can be fabricated and assembled on site. This reduces the transportation cost of materials.

ii. There is less wastage of raw materials – contrary to traditional methods of producing construction materials such as moulding, 3D printing produces the materials in desired dimensions and shapes.
iii. 3D printing technology provides an opportunity to create complex and exciting shapes of buildings and structures. For instance, it is easy to create naturally inspired designs with this technology.
iv. There is reduced manpower and hence the cost of labour.
v. The cost of mass customization is lower with 3D printing since the same CAD design can be reproduced as many times as possible.
vi. 3D printing enhances safety and reduction in injuries at construction sites, especially where special equipment would have otherwise been utilized.

The use of 3D printing in the modern construction industry cannot be achieved successfully without developing the necessary printable materials. Although there are several 3D printing technologies, fused deposition technology, which uses polymeric filaments (materials), has been the most advanced [5]. The fact that it uses polymer-based filaments makes it very suitable for the construction industry since polymers, composites, and ceramics are the most used materials in this industry. The focus should be on developing sustainable 3D printable polymer composites for the construction industry.

Generally, some of the polymer materials which can be used in 3D printing for the construction industry include the following:

- Polystyrene: This material is traditionally manufactured through moulding techniques for the production of construction components such as window panes and sheets. Luckily, fused deposition modelling 3D printers have been developed for this material. However, its application in the construction industry (3D printing) is limited. The 3D printed polystyrene can be utilized for mild structural applications such as partitioning, framework, and interior décor. It is important to note that the properties of the pure polymer should be enhanced through reinforcements with natural fibres and particles (polystyrene polymer composite) for effective application in the industry.
- Polyurethane: The material is very suitable for flexible applications such as flexible foam used in upholstery. In the construction industry, it has been used as a sealant, insulators, and for fire-stopping. It can also be used as adhesive in the construction industry. It is possible to produce polyurethane wood composites that can be used

in structural applications in buildings. The material has been 3D printed, especially through stereolithography (SLS) and therefore its applications in the construction industry can be an exciting venture for better buildings/structures in the future.

- Cement/silica-based composites: The 3D printing for cementitious composites for structural applications has been on the rise [6 , 7]. For instance, [8] have demonstrated the possibility of using fibre-reinforced cement composite for concrete 3D printing. Although the use of 3D printed cement in the industry has grown, there is a need to continuously develop more materials especially polymer-cement composites for 3D printing and construction industries. Such materials shall be important for the development of self-healing materials for flexible buildings and structures.
- Other materials with potential for 3D printing for the construction industry include acrylonitrile butadiene styrene (ABS), polylactic acid (PLA), wood-based materials, polyamide, and nylon. These materials have successfully been 3D printed using Fused Deposition Modeling (FDM) systems, although their applications in the construction industry have not been explored as per the knowledge of the authors of this book. The materials can further be investigated for polymer-based composites for 3D printing of construction materials in the future.

Direct 3D printing of buildings and structures can save on cost and time and enhance designs. As mentioned, it is also possible to construct bioinspired or naturally inspired structures. In this section, some of the 3D-printed buildings and structures are illustrated as case studies (Figures 6.1–6.6). With the increasing need for aesthetics and complex shapes of structures,

FIGURE 6.1 Comparisons of construction methods: (a) conventional and (b) 3D printing technologies (reused from [9] with permission from Elsevier Ltd).

FIGURE 6.2 Building erected through 3D printing by Apis Cor in Dubai: (a) during and (b) after completion of 3D printing (reused from [9] with permission from Elsevier Ltd).

FIGURE 6.3 Typical examples of application of 3D printing: (a) bus stop (b) Trabeculae pavilion, and (c) Gemert bicycle bridge (reused from [9] with permission from Elsevier Ltd).

3D printing technology will no doubt continue to expand in the construction sector.

6.3 EMERGING COMPOSITE MATERIALS

The need to safeguard the environment and reduce disposal dictates the development of new composite materials for construction and related industries. Here, a few of the emerging materials for related applications are highlighted:

i. Keratin starch bio-composite film from waste chicken feather and ginger starch: A recent publication in *Nature Scientific Reports* by

FIGURE 6.4 More examples of 3D printed structures (obtained from [10] with permission from Elsevier Ltd).

[11] illustrated the development of such composite at the Landmark University in Nigeria. The bio-composite was shown to exhibit attractive stability in water. Further studies are necessary for these composites to evaluate their applications in the construction sector.

ii. Lightweight banana fibre paper bricks: This is an interesting composite developed by [12] and it was shown to have potential applications in the construction industry as a walling material. Further related fibre and paper-based bricks possess the potential for lightweight bricks in the future industry. There is a need however to explore the enhancement of their structural strength.

iii. PVC/bamboo/coconut shell hybrid composite: This has been recently developed by [13] and published in *Nature Scientific Reports* in 2021. The authors indicated that the composite has the potential for applications in the construction sector.

iv. Textile reinforced hybrid composites: Hybrid composites reinforced with waste textile materials have been recently developed and evaluated for mechanical and chemical properties by [14]. The composite is an emerging material for the construction industry.

v. Regolith-based magnesium oxychloride composites: These materials have been recently developed by [15] for advanced performance in the construction industry.

FIGURE 6.5 Examples of bioinspired buildings (obtained from [10] with permission from Elsevier Ltd).

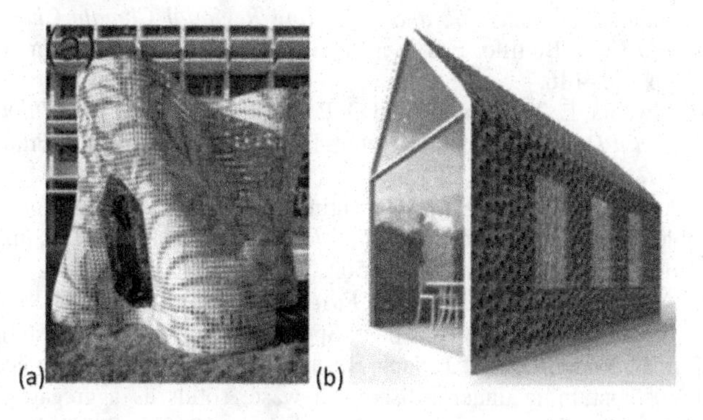

FIGURE 6.6 Examples of 3D printed structures: (a) bloom and (b) shed (obtained from [10] with permission from Elsevier Ltd).

6.4 SUMMARY

In this chapter, an outlook of the future of composite materials in the construction sector has been presented. It is noted that the need for safety, comfort, security, and aesthetics in buildings and structures has pushed the adoption of modern technologies and materials in the construction sector. Additionally, the concept of Industry 4.0 society requires improved standards and quality of processes and systems. To achieve fully the state of digitalization and future construction need, there has to be the development of better and high-performance materials, and composites' role cannot be overlooked in this case. The importance of 3D printing in the construction sector (in modern society and the future) has been highlighted, and it is agreeable that the buildings for the future shall be fabricated through 3D printing. Finally, examples of emerging composite materials for the sector have been provided, and it is no doubt that there is excitement among the researchers in the industry to develop new and enhanced composite materials.

REFERENCES

1. Lucintel. *Management Consulting, Market Research Company, Market Research Firms.* [Online]. Available: https://www.lucintel.com/ (accessed June 27, 2021).
2. J. Johns and M. Fedeski, "Adapting building construction to the effects of climate change," in *Detecting and Modelling Regional Climate Change*, M. B. India and D. L. Bonillo, Eds. Berlin, Heidelberg: Springer Berlin Heidelberg, 2001, pp. 605–616.
3. F. M. Mwema, E. T. Akinlabi, and O. P. Oladijo, *Sputtered Thin Films: Theory and Fractal Descriptions.* Boca Raton, FL: CRC Press/Taylor & Francis Group LLC, 2021.
4. M. Sakin and Y. C. Kiroglu, "3D printing of buildings: construction of the sustainable houses of the future by BIM," *Energy Procedia*, vol. 134, pp. 702–711 2017, doi: 10.1016/j.egypro.2017.09.562.
5. F. M. Mwema and E. T. Akinlabi, *Fused Deposition Modeling: Strategies fo Quality Enhancement*, 1st ed. Cham: Springer International Publishing, 2020.
6. G. Bai, L. Wang, G. Ma, J. Sanjayan, and M. Bai, "3D printing eco-friendl concrete containing under-utilised and waste solids as aggregates," *Cemen and Concrete Composites*, vol. 120, no. 1, p. 104037, 2021, doi: 10.1016/ cemconcomp.2021.104037.

7. J. Xiao,G. Ji, Y. Zhang, G. Ma, V. Mechtcherine, J. Pan, L. Wang, T. Ding, Z. Duan, and S. Du. "Large-scale 3D printing concrete technology: current status and future opportunities," *Cement and Concrete Composites*, vol. 122, p. 104115, 2021, doi: 10.1016/j.cemconcomp.2021.104115.
8. J. Sun, F. Aslani, J. Lu, L. Wang, Y. Huang, and G. Ma, "Fibre-reinforced lightweight engineered cementitious composites for 3D concrete printing," *Ceramics International*, vol. 132, no. 8, p. 88, 2021, doi: 10.1016/j.ceramint.2021.06.124.
9. S. Pessoa, A. S. Guimarães, S. S. Lucas, and N. Simões, "3D printing in the construction industry – a systematic review of the thermal performance in buildings," *Renewable and Sustainable Energy Reviews*, vol. 141, p. 110794, 2021, doi: 10.1016/j.rser.2021.110794.
10. A. Du Plessis, A. J. Babafemi, S. C. Paul, B. Panda, J. P. Tran, and C. Broeckhoven, "Biomimicry for 3D concrete printing: a review and perspective," *Additive Manufacturing*, vol. 38, no. 2016, p. 101823, 2021, doi: 10.1016/j.addma.2020.101823.
11. O. M. Oluba, C. F. Obi, O. B. Akpor, S. I. Ojeaburu, F. D. Ogunrotimi, A. A. Adediran, and M. Oki. "Fabrication and characterization of keratin starch biocomposite film from chicken feather waste and ginger starch," *Scientific Reports*, vol. 11, no. 1, p. 8768, 2021, doi: 10.1038/s41598-021-88002-3.
12. A. A. Akinwande, A. A. Adediran, O. A. Balogun, O. S. Olusoju, and O. S. Adesina, "Influence of alkaline modification on selected properties of banana fiber paperbricks," *Scientific Reports*, vol. 11, no. 1, p. 5793, 2021, doi: 10.1038/s41598-021-85106-8.
13. A. A. Adediran, A. A. Akinwande, O. A. Balogun, O. S. Olasoju, and O. S. Adesina, "Experimental evaluation of bamboo fiber/particulate coconut shell hybrid PVC composite," *Scientific Reports*, vol. 11, no. 1, p. 5465, 2021, doi: 10.1038/s41598-021-85038-3.
14. Z. Kamble and B. K. Behera, "Sustainable hybrid composites reinforced with textile waste for construction and building applications," *Construction and Building Materials*, vol. 284, no. 12, p. 122800, 2021, doi: 10.1016/j.conbuildmat.2021.122800.
15. A.-M. Lauermannová, I. Faltysova, M. Lojka, F. Antončík, D. Sedmidubsky, Z. Pavlik, M. Pavlíko, M. Záleská, A. Pivák, and O. Jankovský. "Regolith-based magnesium oxychloride composites doped by graphene: novel high-performance building materials for lunar constructions," *FlatChem*, vol. 26, no. 2, p. 100234, 2021, doi: 10.1016/j.flatc.2021.100234.

Index

Printed in the United States
by Baker & Taylor Publisher Services

Printed in the United States
by Baker & Taylor Publisher Services